ART ON CHAIRS

2014–2015

PAREDES . PORTUGAL

椅子上的艺术

《casa 国际家居》杂志 编

新星出版社　NEW STAR PRESS

ART ON CHAIRS 2014–2015
椅子上的艺术 2014–2015

项目推动 Promoter
帕雷德斯市政厅
Paredes Municipality

项目总协调 Global Coordination
SETEPÉS

椅子上的艺术项目(中国)总协调
Coordination Art on Chairs Beijing
《casa国际家居》杂志
casa international
伊曼纽尔·巴博萨
Emanuel Barbosa

合作伙伴 Partners
ID+ / 葡萄牙阿威罗大学
ID+ / Aveiro University
马托西纽什ESAD艺术和设计大学
ESAD – College of Art and Design, Matosinhos
《casa国际家居》杂志
casa internacional

支持 Support
北控水务集团
Bewater / BEWG
Investwood
Urbanos
Banema

资金支持 Cofinancing
ON.2
QREN
欧盟 European Union

项目团队 Project Team

项目总协调 Global Coordination
苏珊娜·马克斯
Susana Marques

项目管理 Project Managment
安娜·维埃拉 Ana Vieira

出品 Production
奥尔加·莫雷拉
Olga Moreira

帕雷德斯式的体验 Exhibition "The Experience of Being By Paredes"
策展人 Curators
弗朗西斯科·博瓦德尼奇，瓦斯科·布兰科
Francisco Providência and Vasco Branco

展览设计 Exhibition Design
弗朗西斯科·博瓦德尼奇
Francisco Providência

协作 Collaboration
保罗·勃顾 Paulo Bago D'Uva
列吉亚·阿夫雷科 Lígia Afreixo
马塔·弗瑞格塔 Marta Fragata
亚历山德拉·巴高 Alexandra Bagão

摄影 Photography
马库斯·加西亚·莫雷拉
Marcus Garcia Moreira

二重奏(2+1) Exhibition "Duets (2+1)"
策展人 Curator
何塞·巴特罗 José Bártolo

协调人 Coordination
塞尔吉奥·阿方索
Sérgio Afonso

展览设计 Exhibition Design
若·克鲁兹 João Cruz

摄影 Photography
伊内斯·迪奥伊 Inês D'Orey

出品 Production
安娜·梅德罗斯 Ana Medeiros

椅子游行 Exhibition "Chair Parade"
协调人 Coordination
桑德拉·劳 Sandra Lau

展览设计 Exhibition Design
弗朗西斯科·博瓦德尼奇
Francisco Providência

摄影 Photography
马库斯·加西亚·莫雷拉
Marcus Garcia Moreira

葡式设计 Exhibition "How to pronounce Design in Portuguese?"
项目推动伙伴 Co-promotor
里斯本市政厅 Lisbon Municipality
里斯本艺术设计博物馆 MUDE – Museu do Design e da Moda, Francisco Capelo's Collection

策展人 Curator
芭芭拉·科蒂纽 Bárbara Coutinho

展览设计 Exhibition Design
马里亚诺·皮萨罗 Mariano Piçarra

平面设计 Graphic Design
TVM / Luís Moreira

图片 Images
版权归属作者所有
The rights belongs to the authors

出品 Production
卡塔琳娜·赛德 Catarina Cid

展览设计 Exhibition Design
Guarnição

图册 Catalog
椅子上的艺术 2014–2015
ART ON CHAIRS 2014–2015

编辑协调 Editorial Coordination
SETEPÉS

文字 Texts
芭芭拉·科蒂纽 Bárbara Coutinho
弗朗西斯科·博瓦德尼奇
Francisco Providência
何塞·巴特罗 José Bártolo
罗萨·爱丽丝·布兰科 Rosa Alice Branco
桑德拉·劳 Sandra Lau

翻译 Translation
蔡婧 Cai Jing
Phala
丽塔·阿尔维斯 Rita Gonçalves

平面设计 Graphic Design
马塔·博尔赫斯 Marta Borges

版本 Edition
帕雷德斯市政厅 Paredes Municipality

出版时间 Year of publication
2014

本图书所介绍的是"2014–2015椅子上的艺术"在北京国际设计周上展示的内容。
This catalogue is published in the frame of Art on Chairs 2014–2015 participation at the Beijing Design Week 2014.

序言

"椅子上的艺术"是一项鼓励设计和当代创意、推动行业创新和跨界发展的国际活动。

"椅子上的艺术"由葡萄牙帕雷德斯市政厅发起,该活动旨在将该区打造成一个高端家具制造中心。为此,"椅子上的艺术"集合了创意和生产、文化和经济、本地财富和全球机遇。

北京是"椅子上的艺术"全球巡展的首站,这也是"椅子上的艺术"第一次将展示第一站选择在中国。这场全球家具设计展览会将持续至2015年。

第二届"椅子上的艺术"将从多角度展示葡萄牙设计和帕雷德斯当代家具作品,展览将涉及帕雷德斯家具业的新设计产品、本地学校的创意展示以及过去十年帕雷德斯家具设计回顾。

在2014年北京国际设计周上"椅子上的艺术"包括以下四个板块:

"帕雷德斯式的体验"向葡萄牙设计师发出挑战,要求他们针对高附加值市场设计作品,并通过设计回答一个共同的问题:如何在避免简单的奢华展示下进行设计表达。

"二重奏(2012和2014)"为每一位设计师指定一个著名人物,要求设计师设计的椅子必须充分反映人物的个性特征和形象,在这场对话中设计既是诠释语言也是筑造工具。

"椅子游行"这一教育倡议旨在鼓励对艺术创意表达的审美、批判性观察和实验。一千名学生、一百把椅子和一个挑战:对20世纪和21世纪的葡萄牙平面设计师和插图作家的作品进行创意改造。

"葡式设计"展览试图在全球文化融合的大背景下反思葡萄牙设计DNA存在可能性的问题。

Prologue

Art on Chairs is an international event that promotes design and contemporary creativity as driving forces for innovation in industry and development across the board. An initiative by the Portuguese Municipality of Paredes that seeks to reposition the territory as a reference in the production of high-quality furniture. Art on Chairs brings together creation and production, culture and economy, local assets and global opportunities.

Beijing is the first stage of Art on Chairs' international presentation; the first stop of a route that starts in China (September 2014), is presented in Portugal (Paredes and Lisbon), and culminates in 2015 with an international circuit through the main furniture design fairs. On this second edition, Art on Chairs presents Portuguese design and Paredes' contemporary furniture production on a plural approach. A wide range of perspectives: the production of new design objects produced by Paredes industry; creative interventions from the local educational community and a retrospective on Portuguese furniture design over the last decade of the 21st century.

At Beijing Design Week 2014 Art on Chairs debuts with four exhibitions: The Experience of Being by Paredes, a challenge to Portuguese designers to develop new objects oriented towards high value-added market; these proposals aim at answering the same question: how to interpret the signals that take off the idea of simple luxury demonstration? **Duets** (2012 and 2014), in each duet the designer reflects the identity and material culture of a guest personality in a unique chair designed at his/her image, it's a dialogue where design is both an interpretative language and a constructive tool. **Chair Parade**, an educational initiative to stimulate aesthetic enjoyment, critical observation and experimentation in art creative expression. 1,000 students, 100 seats and a challenge: change them according to the work of Portuguese graphic designers and illustrators of the 20th and 21st centuries. **How to pronounce design in Portuguese?**, an exhibition that brings up the reflection on the possible existence of a DNA of Portuguese design, given the current global culture of the discipline.

目录
Contents

1 葡萄牙经典家具设计来到中国
The Portuguese Timeless Art of Good Furniture Manufacturing Reaches China

5 行业
Industry

6 帕雷德斯式的体验
The Experience of Being by Paredes

62 二重奏
Duets

102 椅子游行
Chair Parade

118 葡萄牙帕雷德斯制造
Made by Paredes

120 葡式设计 2000–2014
How to Pronounce Design in Portuguese? (2000 – 2014)

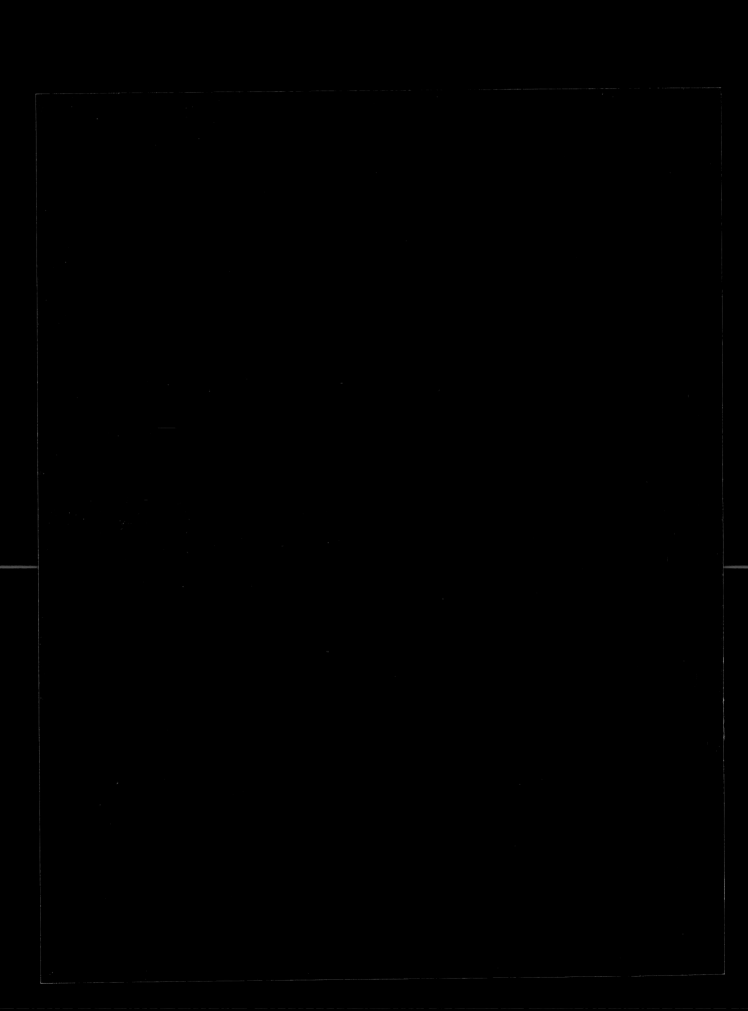

椅子上的艺术"充分展示了帕雷德斯精良的家具生产传统和技艺。800多家来自帕雷德斯的家具制造企业体现出了这个具有高超木工技艺的小镇独一无二的艺术遗产，同时也为这个开放的国度增添更多的历史价值。

葡萄牙有世界上最早的边界线。早在五个世纪前葡萄牙人就首次与中国开展交流，从那时起，葡萄牙文化在世界各地留下印记。

葡萄牙普世和全球化的命运也同样体现在它深深扎根于新兴文化的所在地——非洲和美洲。安哥拉、莫桑比克和巴西都是葡萄牙的兄弟国家，代表着居于世界各地的2.5亿葡语人群。

2012年"椅子上的艺术"可以说是当年最佳欧洲项目，它获得了"RegioStars"大奖——这一由欧洲委员会颁发的最有声望的奖项。2014/2015"椅子上的艺术"将继续遵循帕雷德斯提出的发展战略，即推动家具行业竞争性、向世界展示欧洲最好家具的品质。

第二届"椅子上的艺术"反映艺术、设计和葡萄牙工业的普世性。

这里的家具行业将普世性作为家具设计灵感的来源，所有参与的企业、企业家和设计师成为了这个代表卓越的行业的形象大使。

为实现这个目标，超过2600人投入到设计、创意、制版、试验和认证流程中。通过我们的企业和举办的国际设计和项目大赛，通过本地的教育机构，帕雷德斯向世界展示了她的创意，展示了行业和教育之间所希望达到的基本结合。

"二重奏"是其中最突出的项目。"二重奏"项目是一项社会责任倡议，它是由11名著名设计师按照11个世界知名人士的性格设计的椅子，再由帕雷德斯家具企业按照设计制造实物。葡萄牙总统阿尼巴尔·卡瓦科·席尔瓦、克里斯蒂亚诺·罗纳尔多、何塞·穆里尼奥、卢西亚诺·贝纳通都曾入选"二重奏"名人之列。之后，11把椅子已被拍卖，拍卖所得全部捐赠给联合国难民署。

在2014/2015"椅子上的艺术"中，"二重奏"项目已确认了今年所选定的世界知名人士：欧洲委员会主席巴罗佐、巴西前总统卢拉、前足球球星马拉多纳、电影演员约翰·马尔科维奇以及F1方程式赛车世界冠军手巴顿。

今年的"二重奏"项目将更加国际化，在中国的展出将成为最受关注的焦点之一。

因此，我们宣称"椅子上的艺术"是一项享有崇高国际声誉的独一无二的项目。"椅子上的艺术"充分维护了欧洲设计葡萄牙制造的品质。

谢谢！

塞尔索·费雷拉
帕雷德斯市长

The Portuguese Timeless Art of Good Furniture Manufacturing Reaches China

Art on Chairs is an event that truly reflects the ancient tradition and know-how of good furniture production of Paredes. The unique artistic legacy of a community that masters woodworking is embodied in the almost 800 furniture manufacturing units of Paredes, which further add the value of the history of a nation that opened the World to globalization.

Portugal has the oldest borders in the world, and Portuguese culture has left its traces throughout the world for more than five centuries, when it first started its privileged exchanges with China.

Portugal, with its universalistic and globalizing destiny, was also at the root of new cultures, namely in Africa and America. Angola, Mozambique and Brazil are examples of brother countries that represent a Portuguese-speaking community of about 250 million people worldwide.

Art on Chairs 2012 was considered the year's best European project, having received the "RegioStars" award, the most prestigious award offered by the European Commission. Art on Chairs 2014/2015 follows the strategy defined by Paredes for promoting the competiveness of the furniture sector, aimed at bringing to the world the quality of the best furniture made in Europe.

This second edition of Art on Chairs reflects the universal nature of art, design, and of the Portuguese industry.

Participating companies, entrepreneurs and designers become true ambassadors for this industry of excellence, which uses universality as a true source of inspiration for design.

To reach such a point of excellence, over 2,600 people have worked in design, creation, prototyping, production, testing and certification. We have held creative residences in companies, as well as international design contests and projects wich also included local schools and universities, in a desired and fundamental articulation between the industry and education.

Duets is, however, the most prominent face. Duets is a social responsibility initiative, by which 11 renowned designers have created chairs for 11 international personalities, which were then produced by furniture companies in Paredes. The President of the Portuguese Republic, Mr Aníbal Cavaco Silva; Cristiano Ronaldo; José Mourinho, or Luciano Benetton featured some of the duets. The 11 chairs were then auctioned for the global amount of 111,500 Euros, which were donated to the United Nations High Commissioner for Refugees.

Duets will feature a new edition on Art on Chairs 2014/15, and various personalities have already been confirmed, namely Mr Durão Barroso, the President of the European Commission; Mr Lula da Silva, the former President of Brazil; Diego Maradona, a former football star; Mr John Malkovich, a film actor; and Jenson Button, Formula 1 World Champion.

With this new edition the event shall become truly international, with the exhibition in China being one of its most prominent highlights.

For all these reasons, we state that Art on Chairs is a unique event with international prestige. Moreover, Art on Chairs truly asserts the quality of European design made in Portugal.

Thank you very much.

Celso Ferreira
Mayor of Paredes

行业

将工匠转型为艺术家的过程实在是一件趣事。

——弗朗西斯科·博瓦德尼奇

帕雷德斯家具企业是这一地区经济发展的引擎。这一行业对于葡萄牙品牌的国际化和品质的确立发挥了关键作用对于该地区和葡萄牙具有重要意义。

如今，在这片大约160平方公里的土地上有超过800家木制品制作企业，其中包括保留了18世纪传统工艺的工坊，也有追求高品质的现代企业。生产和技术的结合使得这里的企业可以灵活运用各种现代技术加工木材和相关产品，确立了一大批占据重要市场地位的家具品牌。

帕雷德斯发起的"椅子上的艺术"旨在扩大帕雷德斯的国际知名度，设计和生产最高品质的木质工艺品。这个项目的成功举办离不开所有参与企业的贡献，它们欢迎设计师和策展人进入它们的工厂，进行产品设计和开发。

在应对挑战和执行实验的过程中，这些企业不断突破能力极限，解决难题，充分体现了"椅子上的艺术"所倡导和分享的追求卓越的理念。

Industry

There is a pleasure in making that transforms the workman into an artist.
Francisco Providência

The furniture companies of the Paredes Municipality are the driving force of the region's economy. Vital for the region but also for the country, this industry contributes in a significant way to internationalising and establishing the quality of the Portugal brand.

Today, within approximately 160 km^2, more than 800 wood manufacturing units are concentrated in the municipality of Paredes, covering an extensive range of specialisations from artistic workshops (where traditional techniques dating from the 18th century are preserved), to sophisticated modern industries subject to the highest standards of quality. The concentration of production units and their technical specialisation employ every available contemporary technology dedicated to the processing of wood and its derivatives, establishing an important group of commercial furniture brands within the market.

Paredes Municipality created Art on Chairs with the aim of positioning Paredes in the international furniture market, designing and producing wood artefacts to the highest quality standards. This project was only possible because of the efforts of the participating companies that welcomed the designers and curators into their factories for the development and prototyping of new pieces.

In their response to the challenges set and their experimentation with solutions, these companies have explored new areas of their expertise and their ability to fulfil them, in a commitment to excellence that Art on Chairs shares and celebrates.

BEIJING
China
北京
中国

帕雷德斯式的体验

THE EXPERIENCE OF BEING
BY PAREDES

CURATORS

FRANCISCO PROVIDÊNCIA
DESIGNER, RESEARCHER AT ID+, TEACHER AT AVEIRO UNIVERSITY

VASCO BRANCO
ENGINEER, RESEARCHER AT ID+, TEACHER AT AVEIRO UNIVERSITY

策展人

弗朗西斯科・博瓦德尼奇
设计师，ID+调研员，葡萄牙阿威罗大学讲师

瓦斯科・布兰科
设计师，ID+调研员，葡萄牙阿威罗大学讲师

LUXURY IS A NECESSITY THAT BEGINS WHERE NECESSITY ENDS (...) IT IS NOT THE OPPOSITE OF POVERTY, IT IS THE OPPOSITE OF VULGARITY. LUXURY IS EVERYTHING YOU DON'T SEE.
COCO CHANEL

当生活必需品都具备之后，奢华品就成为了必要，奢华的对立面并不是贫穷，而是粗俗。奢华是你用肉眼无法看到的。

——可可・香奈儿

帕雷德斯式的体验

弗朗西斯科·博瓦德尼奇，瓦斯科·布兰科

奢侈是……吃野天鹅肉

在英国，自从12世纪以来，捕猎野天鹅成为皇家专享的特权，为的是制成正式宴会的佳肴。由于稀少，这种神秘的皇家专享体验便成了一种体现品质和权利的奢侈。

葡萄牙自12世纪立国以来在这片土地上形成的统一文化使得居住在北方的人接触到地中海和阿拉伯文化，比之锡哥特—罗马文化遗产来说，这两种文化更加世俗和富有美感，技术上也更复杂。

繁荣的象征，帕雷德斯（葡萄牙）

奢侈已经成为越来越多的知名品牌成功的商业条件。葡萄牙帕雷德斯特别邀请了10位设计师以奢侈为主题打造原创木制设计。

今天，帕雷德斯地区内有超过600件木质房屋，集中在一片约160平方公里的土地上。其中既有保留了18世纪传统工艺的艺术工坊，也有遵循严格质量标准的现代工业企业。在对奢侈的标准进行解读时，受邀设计师通过每一件作品进行表达，既兼顾物件的功能性，又在形式上浓缩着对于未来的新理念。不仅仅是选材和做工上追求卓越（例如爱马仕和乐途仕的手工制品），更重要的是通过精巧的外形所传递的新理念，高调宣传着对于品质生活的追求。"帕雷德斯式的体验"展览将奢侈的概念从简单的购买力炫耀转化为对于生活本身的成就感。

The Experience of Being by Paredes

Francisco Providência and Vasco Branco

LUXURY IS ... EATING WILD SWAN

In the UK, since the twelfth century, capturing of wild swans has been exclusively a royal privilege; royalty is entitled to the exclusive right of hunting them, in order to serve them at official banquets. The mythic exclusivity of this dining experience produced a sense of luxury, both in quality and in power, guaranteed by its rarity.

Portugal has extensive economic experience associated with luxury. With the cultural unification of the territory deriving from the founding of the nation (12th century) the northern populations came to enjoy a Mediterranean and Arabic culture, more sensual and artistic, and technologically more sophisticated than the Visigoth–Roman heritage.

BY PAREDES (PORTUGAL), UNDER THE SYMBOL OF PROSPERITY

Luxury status, now a commercial condition for the success of the more demanding brands, was the theme that brought together the group of ten designers to create original pieces, designed and built in wood *by Paredes (Portugal)*.

Today, the young municipality of Paredes has more than 600 wood manufacturing units concentrated within approximately 160 km^2, covering an extensive range of specialties from artistic workshops, where traditional techniques from the 18th century are preserved, to sophisticated modern industries subject to the highest quality standards.

The interpretation of the criteria of luxury by the invited designers, expressed in set of pieces in which, together with a functional pretext, forms are explored which encapsulate new ideas for the future. It is not just the use of exceptional techniques and materials (as we see, for example, in handmade products by Hermés or Lottusse) but, above all, new ideas that, in their precise appearance, profess a greater demand for quality of life.

The exhibition "*The experience of being by Paredes*" shifts the idea of luxury, from the simple demonstration of purchasing power, to the fulfilment of life itself.

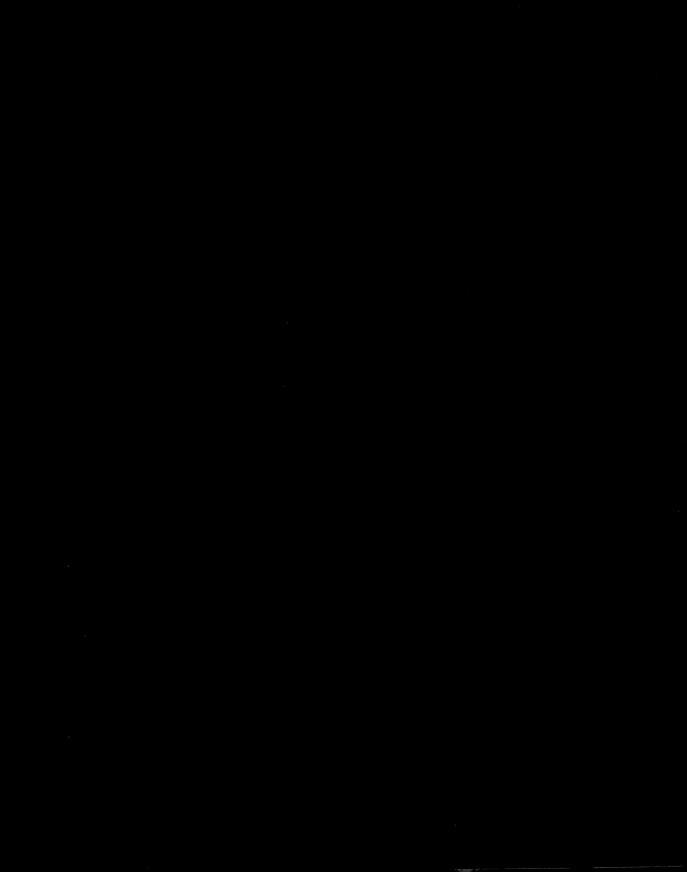

何塞·阿劳若 SUIDENETO	JOSÉ CARVALHO ARAÚJO SUIDENETO
马尔科·索萨·桑托斯 LOUREIRO SANTOS	MARCO SOUSA SANTOS LOUREIRO SANTOS
马尔科·索萨·桑托斯 CAMILA MÓVEIS	MARCO SOUSA SANTOS CAMILA MÓVEIS
马提亚·莱纳 GUALTORRES	MATTHIAS LEHNER GUALTORRES
马提亚·莱纳 MARGEM IDEAL	MATTHIAS LEHNER MARGEM IDEAL
保罗·费雷拉·内维斯 CUNHA MOBILIÁRIO	PAULO FERREIRA NEVES CUNHA MOBILIÁRIO
保罗·费雷拉·内维斯 RABISCOS SENSATOS	PAULO FERREIRA NEVES RABISCOS SENSATOS
佩德罗·卡斯特 MÓVEIS FIALHO	PEDRO KARST MÓVEIS FIALHO
佩德罗·索萨 RONFE CLASSIC	PEDRO SOUSA RONFE CLASSIC
佩德罗·席尔瓦·迪亚斯 A. BRITO	PEDRO SILVA DIAS A. BRITO
佩德罗·席尔瓦·迪亚斯 WOODSPACE	PEDRO SILVA DIAS WOODSPACE
鲁伊·阿尔维斯 SEABRA MÓVEIS	RUI ALVES SEABRA MÓVEIS
鲁伊·阿尔维斯 PELCORTE	RUI ALVES PELCORTE
鲁伊·格拉兹纳 IDEALMÓVEL	RUI GRAZINA IDEALMÓVEL
鲁伊·格拉兹纳 ZAGAS	RUI GRAZINA ZAGAS
鲁伊·维亚纳 CÓMODOS	RUI VIANA CÓMODOS
鲁伊·维亚纳 EVANY ROUSE	RUI VIANA EVANY ROUSE

分享的桌子

何塞·阿劳若
与丽塔·贾丽佐合作

制作：Guarnição，帕雷德斯

A TABLE TO SHARE

DESIGNED BY **JOSÉ CARVALHO ARAÚJO**

IN COLLABORATION WITH RITA GARIZO

PRODUCED BY GUARNIÇÃO

I HAVE LEARNED THAT
TO BE WITH THOSE I LIKE
IS ENOUGH.
WALT WHITMAN

我发现，只要和喜欢的人在一起就足够了。

——沃尔特·惠特曼

"同伴"意味着那些和我们分享面包的人。出于信任,我们和同伴们分享桌子和观点。何塞·阿劳若设计了一张配有四把椅子的桌子。它的寓意是同伴是桌子的基础,没有同伴就没有桌子。

以脆弱的垂直连接支撑桌面的形式表达一种奇异和荒诞,但又不是完全不可能,从而通过这一设计传递出一种真实和承诺。这种对于桌子功能性的超越所揭示的是,通过削弱结构形态的表达体现一种与众不同和奢华感。Guarnição公司以最高质量标准的要求打造了这件作品,精致的漆面使它显得尤其与众不同。

COMPANIONS (from the Latin "cum panis") meaning those with whom we share bread. A sign of great trust, we share the table and ideas with our companions. José Carvalho Araújo has designed a suspended table for four chairs. The metaphor suggests that companions form the basis of the table, and not the other way round.

The fragile vertical connection that supports the tabletop transmits a sense of the improbability, if not the actual impossibility (of sharing), which makes this object a voice of truth and promises. The evident transcending of the functional aspect of the table-object reveals, through the reduced expression of constructional form, its claim to distinction and luxury. Built to the most demanding standards of quality and rigour by the manufacturing company Guarnição, it stands out through the expression of its finishes.

分享的桌子

何塞·阿劳若
与丽塔·贾丽佐合作
制作：Guarnição，帕雷德斯

a table to share

José Carvalho Araújo
in collaboration with Rita Garizo
produced by Guarnição

贝特克（一家售卖饮料喝点心的小餐馆）

何塞·阿劳若
与丽塔·贾丽佐合作

制作：Suideneto，帕雷德斯

BOTECO
(A SIMPLE ESTABLISHMENT THAT SELLS DRINKS AND LIGHT REFRESHMENTS)

DESIGNED BY **JOSÉ CARVALHO ARAÚJO**

IN COLLABORATION WITH RITA GARIZO

PRODUCED BY SUIDENETO

WHAT I AM AFRAID OF IS
YOUR FEAR
WILLIAM SHAKESPEARE

我所害怕的正是你的恐惧。
——威廉·莎士比亚

以材料的厚重对抗对于不确定性的担忧。Mater（母亲）、materia（材质）和matrix（母体）这三个词都源于单词mother（母亲）。母体是生命之源，它给万物提供保护。母亲是永远欢迎我们的家。Boteco桌子的外形和厚度使人联想到母亲的坚强，总是张开双臂欢迎我们，就像在葡萄牙小酒馆里厚重的长凳，所有人一喝上葡萄酒就成了兄弟。由家具制造商Suideneto打造的这款桌子将桌子的厚重与美酒的轻盈相结合，就像房子和冲浪板的对比。这是一张带你飞翔的桌子，一张庆祝世界和平的桌子。

BOTECO Material thickness for confronting the fear of uncertainty. Mater, materia and matrix are Latin words that originate from the word "mother"; the material mother in the origin of life and remembered protection, common to all mammals. Mother is the home that welcomes us. The form and thickness of the Boteco table evokes maternal solidity. It is there to welcome us, resolute and solid, like the long benches that, in the taverns of Portugal, turn all men into brothers when they drink red wine. Produced by the furniture makers Suideneto, the association of the heaviness of the table with the lightness of wine is inevitable, being simultaneously house and surfboard. A table to fly on, or to celebrate world peace, I would say.

贝特克（一家售卖饮料喝点心的小餐馆）

何塞·阿劳若
与丽塔·贾丽佐合作
制作：Suideneto，帕雷德斯

boteco (a simple establishment that sells drinks and light refreshments)

José Carvalho Araújo
in collaboration with Rita Garizo
produced by Suideneto

扶手椅 + 阅读边桌

马尔科 · 索萨 · 桑托斯
与佩德罗 · 佩德罗索合作

扶手椅 | 制作： José Fernando Loureiro dos Santos， 帕雷德斯

阅读边桌 | 制作： Camila Móveis, 帕雷德斯

ARMCHAIR + SIDE TABLE
FOR SOLITARY READERS

DESIGNED BY **MARCO SOUSA SANTOS**

IN COLLABORATION WITH PEDRO JORGE PEDROSO

ARMCHAIR PRODUCED BY JOSÉ FERNANDO LOUREIRO DOS SANTOS, PAREDES

SIDE TABLE PRODUCED BY CAMILA MÓVEIS, PAREDES

THERE IS NO PLEASURE MORE COMPLEX THAN THAT OF THOUGHT.
JORGE LUÍS BORGES (AUTHOR OF THE IMMORTAL)

没有什么乐趣比思想的乐趣更复杂。

——豪尔赫 · 博尔赫斯
（著有《不朽》）

这把由马尔科·索萨·桑托斯设计、José Fernando Loureiro dos Santos公司制作的扶手椅,通过舒适和保护性的外形使人获得读书时独处的快乐。这款舒适的椅子非常适合放在阳台和走廊边,在炎热的夏夜伴着蟋蟀的低鸣,为人们提供一个舒适的阅读空间。特殊的结构设计制造了虚与实、直线和曲面的对比,精致的美国胡桃木雕刻传递着一种视觉的诗意。

本色胡桃木的色彩会随着时间的变化而产生变化:或紫色或蓝色。Camila Móveis公司制作的这张阅读边桌,通过表面的抛光和设计所传递的独特气氛带给使用者难以忘怀的独特享受。

THE ARMCHAIR designed by Marco Sousa Santos and magnificently built by the company José Fernando Loureiro dos Santos, inspires the solitary pleasures of reading through its relaxed and protecting form. It is an armchair designed for flying while sitting back comfortably, and as such, should sit at the edge of a veranda or on a porch, serving the maternal purposes of accommodating reading that, on hot summer nights, is done to the sound of crickets. The existence of the exposed structure that is transformed into a supporting surface, creates a tension in the design between solids and voids, lines and surfaces, and that are themselves visual poetry sculpted in American Walnut.

The brown of the walnut, which at times tends to purple or blue is, in the slenderness of its structure and the clarity of its architecture, an example of the intelligibility provided by design as against the hermetic opacity of technology. The sensuality of the polished touch of the wood and the intense aroma of the armchair and table built by the Camila Móveis company, will be forever engrained in the memory of those who abandon themselves to its care.

扶手椅 + 阅读边桌

马尔科·索萨·桑托斯
和佩德罗·佩德罗索
扶手椅｜制作：José Fernando Loureiro
dos Santos，帕雷德斯
阅读边桌｜制作：Camila Móveis，
帕雷德斯

armchair + side table for solitary readers

Marco Sousa Santos
in collaboration with Pedro Jorge Pedroso
armchair produced by José Fernando
Loureiro dos Santos, Paredes
side table produced by Camila Móveis, Paredes

家庭餐桌+椅子

马提亚·莱纳

制作：Gualtorres，帕雷德斯 + Margem Ideal，帕雷德斯

FAMILY TABLE + CHAIR

DESIGNED BY **MATTHIAS LEHNER**

PRODUCED BY GUALTORRES, PAREDES + MARGEM IDEAL, PAREDES

TIME, THAT SHEER RESTLESSNESS OF LIFE AND ITS ABSOLUTE AND INHERENT PROCESS OF DIFFERENTIATION.
GEORG HEGEL (PHENOMENOLOGY OF SPIRIT)

时间是永不停息的生命，是分化的绝对和内在的过程。
——**黑格尔**（《精神现象学》）

在当前社会快节奏生活的主基调下，人与人之间交流的机会变得越来越少。因此，马提亚·莱纳提出可将时间中区分出一类作为真正的奢侈：与家人和朋友享受美食的时间。因此，家庭餐桌和与之相配的椅子成了当代居住中万能的交流中心。这个项目的名字"猛犸象"揭示出这个设计的灵感来自仿生学。通过Gualtorres公司提供的技术指导和工业制造，在大象骨架上覆盖上等皮革，从而实现整体的和谐统一。设计师还表示，这款设计遵从了手工艺传统，在过去，不同文化背景下的人们都会从自然中寻找外形概念并将之融入设计中。因此，这些形象概念很容易被世界各地的人们辨认出来从而产生共鸣。

TO COUNTER OUR HECTIC DAILY LIVES IN WHICH WE NOW PASS one another without seeing each other, Matthias Lehner proposes to reclassify time as a true luxury: time enjoyed with family and friends over a good meal.

Therefore the family table – and the chairs that surround it – become the versatile and communicative centrality of contemporary dwelling.

The name of this project – Mammoth – reveals its bionic inspiration that, technically supported by the consulting and industrial manufacturing of Gualtorres, is translated into an elegant skeletal structure covered by fine leather that organically harmonizes the entire set of pieces. According to Matthias Lehner, this design continues the lineage of artisanal tradition, which has always and in different cultural contexts, found formal concepts in Nature for building objects. Forms that, for this reason, can aspire to be recognized globally and connect directly to our collective conscience.

家庭餐桌+椅子

马提亚·莱纳
制作：Gualtorres，帕雷德斯 +
Margem Ideal，帕雷德斯

family table + chair

Matthias Lehner
produced by Gualtorres, Paredes + Margem Ideal, Paredes

多用途支架+椅子

保罗·费雷拉·内维斯

制作：Rabiscos Sensatos + Cunha Mobiliário，帕雷德斯

MULTIPURPOSE SUPPORT + CHAIR

DESIGNED BY PAULO FERREIRA NEVES

PRODUCED BY RABISCOS SENSATOS, PAREDES AND CUNHA MOBILIÁRIO, PAREDES

AESTHETIC PLEASURE MUST BE AN INTELLIGENT PLEASURE.
JOSÉ ORTEGA Y GASSET (AUTHOR OF THE DEHUMANIZATION OF ART)

审美享受一定是理智的享受。
——何塞·奥特嘉·加塞特
（著有《艺术的非人化》）

桌子和椅子是一组配对。但在Cunha Mobiliário公司制作这把椅子时，另一个重要因素也加入了这个配对，即采用实木。作品不断探索各种独一无二的质地、颜色和气味，给予人独特的触摸感受，展现出一种自然的奢华。该设计通过细节强调传统柜子制作的构造技巧，探索结构和方法上的突破以期追忆那个传统技艺盛行的时代。但是，设计师在配件设计上强调的简洁又为这件作品带来了现代感。这种新旧的结合赋予作品以永恒感，从而增加了作品的使用寿命。

IN THE PAIRING OF A DESK – which can also be a tray for a side table – together with a chair, an exceptional partner was found in Cunha Mobiliário, for creating it in solid wood. Exploring the uniqueness of textures, colour and smells, these artefacts reveal the unique identity that surface provides to the touch, an inimitable expression of natural luxury. The design emphasizes in its details the ancient constructional techniques of cabinet making, exploring structure and methods that recall a time of age-old traditional techniques. However, Paulo Neves reintroduces the contemporary nature of the design with the visual simplicity of its components, establishing with old techniques, new and exuberant relationships that lend timelessness to the artefacts, extending their longevity and use.

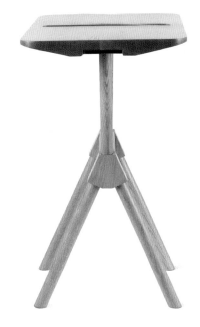

多用途支架+椅子

保罗·费雷拉·内维斯
制作：Rabiscos Sensatos +
Cunha Mobiliário，帕雷德斯

multipurpose support + chair

Paulo Ferreira Neves
produced by Rabiscos Sensatos, Paredes +
Cunha Mobiliário, Paredes

酒吧椅

佩德罗·卡斯特

制作：Móveis Fialho，帕雷德斯

BAR CHAIR TO SIT ON AND STAND OUT

DESIGNED BY PEDRO KARST

PRODUCED BY MÓVEIS FIALHO, PAREDES

MAN IS A CREATION OF DESIRE, NOT A CREATION OF NECESSITY.
GASTON BACHELARD (THE PSYCHOANALYSIS OF FIRE)

男人是欲望的产物，而非必然的产物。

——加斯东·巴什拉

（著有《火的精神分析》）

佩德罗·卡斯特在帕雷德斯一个名为Fialho的工坊找到了属于自己的天地，在这里他可以按照自己的想法设计和打造整个世界。他使用18世纪流传至今的古老木雕技术，将木块雕刻成塑像。它们的存在传递着一种不妥协的自信，仿佛可以暂停时间的流动，质疑世界的脚步。它们履行着自己的社会责任，告诉人们，只要一心一意地遵循这些工艺技巧，你就可以突破所有极限。卡斯特的椅子融合人工和自然，超越了时空的限制带来巴拉克的美感，仿佛上帝之手带给人安宁和幸福。我们可以说这是一把无所畏惧的椅子。

PEDRO KARST FOUND IN PAREDES, and in the crafts workshop of Fialho, a place to rebuild the world, and design it with his own hands. The result are sculptures carved in blocks of wood, constructed hand-in-hand with the techniques of ancient woodcarvers, master craftsmen who seem to have lived since the 18th century unaffected by the constraints of time. Their objects are therefore machines to halt time, which in their uncompromising physical assertiveness, question the progress of the world, and fulfil their social duty to show that it is through the single minded discipline of technique that man can hope to exceed all limits. Karst's chair imposes itself through its form on the space and by its process over time, evokes other times and other baroque spaces in which artificial form seems natural and the technical mastery of nature brings the security and happiness to man that comes from the gratitude of the gods. We might say that this is a chair that asserts itself against fear.

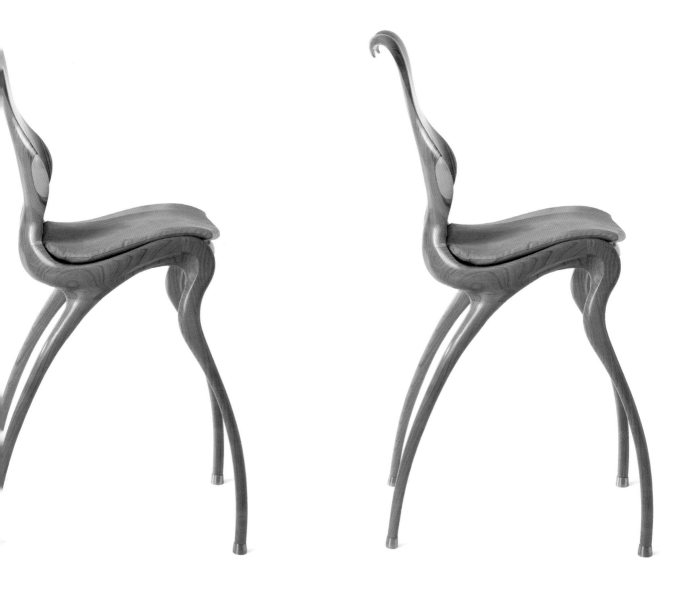

酒吧椅

佩德罗·卡斯特
制作：Móveis Fialho，帕雷德斯

bar chair to sit on and stand out

Pedro Karst
produced by Móveis Fialho, Paredes

小鱼柜子

佩德罗·苏萨

制作：Ronfe Classic，帕雷德斯

CABINET RESEMBLING OF A FISH

DESIGNED BY **PEDRO SOUSA**

PRODUCED BY RONFE CLASSIC, PAREDES

BEAUTY IS BUT
THE PROMISE OF
HAPPINESS.
**STENDHAL (AUTHOR
OF ON LOVE)**

美是幸福的承诺。

——司汤达（著有《论爱情》）

佩德罗·索萨以形式讲述故事，正如《天方夜谭》中的苏丹新娘谢赫拉莎德通过讲述故事分散了人们的注意力，推迟了死亡的到来。他的设计总是包含多重功能，在内容和容器、体积和外表、身体与四肢之间产生一种矛盾的复杂性，使其设计本身具有特殊的吸引力。索萨再次使用葡萄牙传统设计，充分发挥Ronfe Classic公司精于木材表面镶嵌的优势，在表面加工过程中使用了镶嵌技艺。同时使用来自劳沙山区的片岩装饰柜子表面，使人不禁联想到传统印度葡萄牙风格的柜子和桌子。小鱼柜子将我们领向了另一片海域。

STORIES — Pedro Sousa works with form as a pretext to create other narratives, to make stories that, like the legendary Persian king Scheherazade, delay death by distracting us from it. In his work there are always a multitude of functions, producing a paradoxical complexity between content and container, between volume and surface, between body and legs, which lend a strange attraction to the object. Exploring the techniques of surface finishing with marquetry, the speciality of Ronfe Classic, which is renowned for the quality of its wood inlaying, Sousa redoubles the dimensional limits of the form, reusing old designs from Portugal's material culture. The scales of schist that clad or cover the mountain architecture of the Lousã region provide colour and a skin to the chest that evokes the old Indo-Portuguese style of cabinets and desks. In the minutiae of the finishing work, the difference is hidden and the possibility of opposing the determinism of reality is revealed. A fish cabinet carries us to other seas.

小鱼柜子

佩德罗·索萨
制作：Ronfe Classic，帕雷德斯

cabinet resembling of a fish

Pedro Sousa
produced by Ronfe Classic, Paredes

E1P + T1P 交流的桌子

佩德罗·席尔瓦·迪亚斯
与佩德罗·萨尔加多合作

制作：A. Brito和Woodspace，帕雷德斯

E1P DESK AND T1P, PLACE FOR COMMUNICATION

DESIGNED BY PEDRO SILVA DIAS

IN COLLABORATION WITH PEDRO JORGE SALGADO PRODUCED BY A. BRITO AND BY WOODSPACE, PAREDES

BECAUSE OF THE WORD, MAN IS A METAPHOR OF HIMSELF.
OCTÁVIO PAZ (AUTHOR OF THE BOW AND THE LYRE)

因为语言，人是自己的隐喻。
——奥克塔维奥·帕斯
（著有《弓和里拉琴》）

设计就是简化形式，体现本质。形式的精髓则是基于设计者本人的生活。设计者就像是一台生产机器，一件集合知识、力量和主体性的设备（福柯语）。设计者将不同形式结合起来，产生新的组合。设计师将自己的奇思妙想实物化，引入新的叙述方法。席尔瓦·迪亚斯使用了各种技术设备阐述人的寓意，A.Brito公司在制作桌子时使用的椴木能够带来柔软的手感，而另一个由Wodspace公司打造的T1P桌子则使用了雪橇底座，体现一种稳固和自主。

TO DESIGN IS TO REDUCE FORM INTO ITS STYLISTIC ESSENCE and the essence of form is based on the life of the designer himself. The designer is a production machine conditioned by his existence, a device (for knowledge, power and subjectivity according to Foucault). The designer correlates forms and produces new combinations, forms that contain the truth within themselves, materializing ideas that break the storyline to introduce new rules of enunciation. Silva Dias's forms are post-constructivist, maximalist, as round and as absolute as they are phytomorphic and fragile, suggesting a plant-like nature. When designing a contemporary desk for the A. Brito factory to produce with the soft touch of satinwood, Silva Dias summons the technical device with which man enunciated Man himself. In a second prototype designed by Woodspace, the T1P desk gains stability and autonomy from its sled base.

E1P + T1P 交流的桌子

佩德罗·席尔瓦·迪亚斯
与佩德罗·萨尔加多合作
制作：A. Brito和Woodspace，帕雷德斯

E1P desk and T1P, place for communication

Pedro Silva Dias
in collaboration with Pedro Jorge Salgado
produced by A. Brito and by Woodspace, Paredes

100%纯羊毛沙发

鲁伊·阿尔维斯
与安德鲁·阿拉合作

制作：António Seabra Móveis,
帕雷德斯（木材）+ Pelcorte,
帕雷德斯（垫衬物）

SOFA 100% PURE WOOL

DESIGNED BY **RUI ALVES**

IN COLLABORATION WITH
ANDRÉ ARAÚJO

PRODUCED BY ANTÓNIO SEABRA
MÓVEIS (WOOD) + PELCORTE
(UPHOLSTERY), PAREDES

SENSATIONS ARE THE
DETAILS THAT BUILD
UP THE STORIES OF
OUR LIVES.
OSCAR WILDE

感觉构成了我们生活故事的细节。

——奥斯卡·王尔德

鲁伊·阿尔维斯推崇真实的感受，强调身体是过去各种关系的依托，是与这个世界沟通的中介。他的设计为物质文化作出了贡献，换句话说，为母性文化作出了贡献，以设计庆祝现实。对于鲁伊·阿尔维斯来说，真正重要的是过去的东西，那些深植于人们日常生活中的东西，那些我们触摸得到和与之互动的东西。因此，他设计了这款羊毛沙发（羊毛恐怕是人类最早使用的天然高分子材料），Pelcorte公司丰富的制作经验保证了它的绝对舒适。如果说鲁伊·阿尔维斯从手工艺传统中继承了一样东西，那就是清楚地了解到各种限制，并把限制看做是可能性。因此，他的设计中没有什么是随意的，每个细节都经过深思熟虑，因而都具有可操作性。沙发的框架是在António Seabra公司的工坊里加工而成的，充分见证了设计师将古今知识融会贯通，而所有的真理都藏在了细节之中。

100% WOOL Rui Alves, in praise of genuine experiences, emphasizes the body as a receptacle of past relationships, as a home and as a mediator in communication with the world. His design contributes to material culture, or in other words, to maternal culture (Latin mater, the origin of "mother" and "matter"), assigning design to the celebration of reality. For Rui Alves, the truly important things are the banal, those that intrude in the domestic life of people, the things we touch and with which we interact. Therefore, he has designed an alcove sofa that is also a crib because it is made from padded and enveloping wool (perhaps the oldest natural polymer used by humans), giving it a comfort that only Pelcorte's vast experience could guarantee. If there is one piece of knowledge that Rui Alves has inherited from the tradition of craftsmanship, it is a clear understanding of the limits, and of the limit as possibility. For this reason nothing in his design is left to chance; everything is thought out so that everything can be implemented. Constructive, I would say, but constructive in that it substitutes expression for the humble service of life. The structure, sculpted in the workshop of António Seabra, bears witness to this knowledge that is ancient and future; that the truth is in the details.

100%纯羊毛沙发

鲁伊·阿尔维斯
与安德鲁·阿拉合作
制作：António Seabra Móveis,
帕雷德斯（木材）+ Pelcorte,
帕雷德斯（垫衬物）

sofa 100% pure wool

Rui Alves
in collaboration with André Araújo
produced by António Seabra Móveis (wood)
+ Pelcorte (upholstery), Paredes

屏风 + 带暗格的柜子

鲁伊 · 格拉兹纳
与迪奥戈 · 弗里亚斯合作

制作：Idealmóvel和Zagas，帕雷德斯

FOLDING SCREEN AND CABINET WITH HIDDEN COMPARTMENTS

DESIGNED BY **RUI GRAZINA**

IN COLLABORATION WITH DESIGNER DIOGO FRIAS

PRODUCED BY IDEALMÓVEL AND BY ZAGAS, PAREDES

THE TRUTH ABOUT A MAN LIES FIRST AND FOREMOST IN WHAT HE HIDES.
**ANDRÉ MALRAUX
(ANTI-MEMOIRS)**

一个男人的真相最重要的在于他所隐藏的东西。
——安德烈 · 马尔罗
（《反回忆录》）

鲁伊·格拉兹纳设计的这款屏风既能隔开空间,也能将空间连接在一起。不透明的屏风连接着外部的美景和内部的险恶,如同被道德所分隔的不同世界。各种门盒、窗盒和抽屉的安排,构成了充满可能性的架子,而架上的空间则可以放置各种古董。Zagas公司继承了古老的细木工技术、设计和精加工技艺却从不声张,采用实木打造的这些盒子迷宫里珍藏着不愿被遗忘的秘密。这些格拉兹纳的作品因其隐匿的特性,可以看做是私密的物件,呈现了存在主义的秘密,即手眼合一。"看——触"是对设计作为文化调和原则的比喻。格拉兹纳这款由Idealmóvel公司打造的作品,试图通过一组架子将人们引入记忆的迷宫。

THIS SCREEN by Rui Grazina, in its power to separate also unites. It unites the scenic to the "obscenic" (or obscene), through an opacity that betrays the furtive glance; that brings together worlds separated by morality. Its door-boxes, window-boxes, drawer-boxes, make up shelves of possibilities, shelves of empty spaces that could be filled by the archaeology of encounters. These mazes of boxes in solid wood, built by Zagas, conceal ancestral knowledge of joinery, design and finishing, celebrated in a chest in which the secrets that life does not want to forget are hidden. These works of Rui Grazina can be understood as objects of intimacy because of their quality of concealment, contributing to the existential secret of bodies that connect and separate by a protocol that makes the eye and hand a single organ. See-feel is the very metaphor of design as a discipline of cultural mediation.

The design that Grazina asked Idealmóvel to build was a set of shelves for getting lost in the maze of memories.

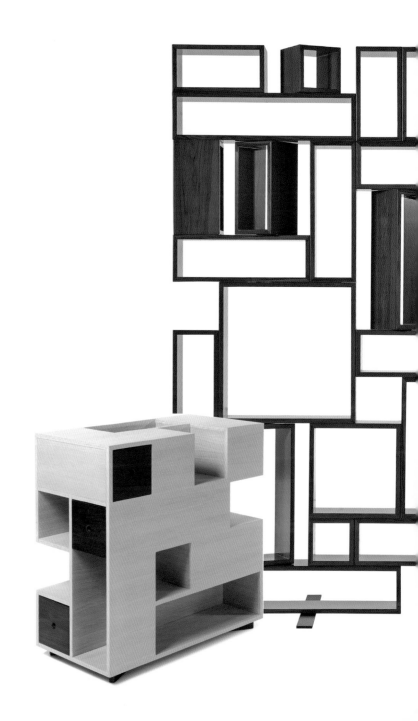

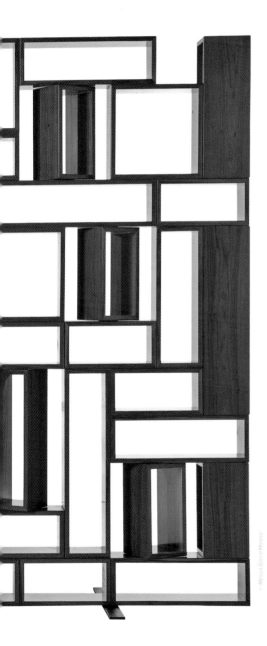

屏风 + 带暗格的柜子

鲁伊·格拉兹纳
与迪奥戈·弗里亚斯合作
制作：Idealmóvel和Zagas，帕雷德斯

folding screen and cabinet with hidden compartments

Rui Grazina
in collaboration with Designer Diogo Frias
produced by Idealmóvel and by Zagas, Paredes

用于游戏的仿生学边桌

鲁伊·维亚纳
与卢·福尔南德斯合作

制作：Cómodos mobiliário，帕雷德斯

BIONIC SIDEBOARD FOR GAMING

DESIGNED BY **RUI VIANA**

IN COLLABORATION WITH
LUÍS FERNANDES

PRODUCED BY CÓMODOS MOBILIÁRIO,
PAREDES

JOY IS THE PLEASURE
OF THE CREATIVE ACT.
KONRAD LORENZ

欢乐就是创造性活动带来的享受。

——康拉德·洛伦兹

鲁伊·维亚纳使用各种技术和材质打造有趣的物件：作品的设计出乎意料又显得毫无功能性可言，成了休闲娱乐的大背景。在维亚纳的设计里，不透明的盒子让位于透明的百叶窗。Cómodos Mobiliário公司熟练地采用六边形磨具打造这款桌子，外形中巧妙融入生物学的理念，展现出一种几何的力量，使人联想到昆虫为了长期生存而建造的巢穴。维亚纳的项目体现了"仿生学"，他喜欢从世界万物中寻找他欣赏的形态，运用到自己的设计中。突破决定论的限制，你会发现新的机会。

BIONIC Rui Viana starts with techniques and materials to construct playful objects; objects that serve as a greater pretext for fun while open to the unexpected and non-functional. In Viana's piece the opacity of the box gives way to the transparency of the shutters, in a game that resembles the pixelated map of artificial images oscillating cell by cell, between opening and closing the opacity of the surface onto the dark interior of the box. Built from hexagonal modules, deftly constructed by Cómodos Mobiliário, the surface reveals the geometric strength of the form converging with gentle biological purpose, bringing to mind colonies organized by insects, designed to better survive over time. Viana's project is bionic, which means that he appropriates the forms he sees throughout the great story of life, using them as pieces of a game that transcends the design's intention. Opening them at random, you see in mistakes the virtue of accessing progress. At the limits of determinism, chance opens onto the new.

用于游戏的仿生学边桌

鲁伊·维亚纳
与卢·福尔南德斯合作
制作：Cómodos mobiliário，帕雷德斯

bionic sideboard for gaming

Rui Viana
in collaboration with Luís Fernandes
produced by Cómodos mobiliário, Paredes

存放番石榴酱的柜子

鲁伊·维亚纳
与卢·福尔南德斯合作

制作：Evany Rouse，帕雷德斯

CABINET FOR STORING GUAVA PASTE

DESIGNED BY **RUI VIANA**

IN COLLABORATION WITH LUÍS FERNANDES

PRODUCED BY EVANY ROUSE, PAREDES

THE MORAL OF ART LIES
IN ITS OWN BEAUTY
GUSTAVE FLAUBERT

艺术的美德在于它的美丽。
——古斯塔夫·福楼拜

如果艺术的美德在于它的美丽，那么艺术将没有美德，因为有了美丽我们就不需要美德了。但是什么是美丽？是什么迷惑了我们？是什么解放了我们？鲁伊·维亚纳的柜子作了装饰性的暗喻。木质纺锤的安装试图最大限度保留情感，Evany Rouse公司的处理方式就是如此，努力营造一种异域和夸张的氛围。使用笼子隔开内外的做法，仿佛来自印度洋、太平洋或者中国。来自海上的季风和雨水带来了各种色彩和气味，带来了佛教的传播，这些都是不能关在笼子里的。这是一款存放番石榴酱或者亚速尔奶酪的柜子。

IF THE MORALITY OF ART LIES IN ITS OWN BEAUTY, then art would have no morality – with beauty we would also not need morality. But what is beauty? That which captivates us, or sets us free? Rui Viana's cabinet is an ornamental metaphor. An installation of wood spindles that, in its rudimentary technique and material truth, seeks to retain maximum emotion, as opposed to minimalism; this is how Evany Rouse approached it, exploring exotic and excessive environments. In its cage-like transparency, it seems to come from the hot and humid countries that are visited when travelling from the Indian Ocean to the Pacific, en route to China. Honeyed cultures, full of colours and aromas, Buddhists, where the wind that brings the monsoon and rain cannot be caged. A cabinet for storing guava paste and *queijo das ilhas* (cheese from the Azores).

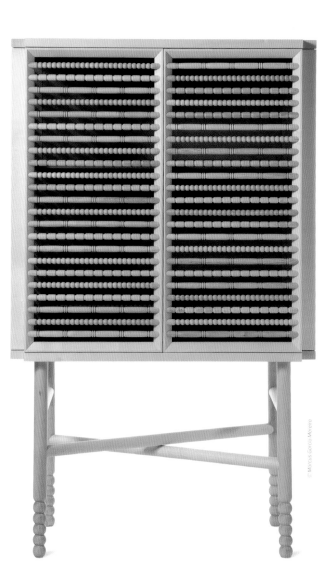

存放番石榴酱的柜子

鲁伊·维亚纳
与卢·福尔南德斯合作
制作：Evany Rouse，帕雷德斯

cabinet for storing guava paste

Rui Viana
in collaboration with Luís Fernandes
produced by Evany Rouse, Paredes

二重奏

DUETS

CURATOR

JOSÉ BÁRTOLO
TEACHER AND COORDINATOR
AT ESAD, MATOSINHOS
ESAD – COLLEGE OF ART AND
DESIGN, MATOSINHOS

策展人

何塞·巴特罗
葡萄牙马托西纽什
ESAD艺术和设计大学
讲师和协调人

DIFFERENT WAYS
OF UNDERSTANDING
DESIGN AND
INTERPRETING THE
PERSONALITIES,
PORTRAYING THEM IN A
BOTH SURPRISING AND
SUITABLE MANNER.
MARIA MILANO

通过不同方式认识设计和诠释人物，这些方式令人感到惊讶又恰到好处。

——玛丽亚·米兰

二重奏

何塞·巴特罗

"二重奏"是一个有着强烈社会责任感和人文关怀的项目。它关注于椅子设计、生产和使用的多种可能性，这种可能性关乎独家设计和工业生产之间的关系，也关乎其文化、经济和技术层面的结果（限量生产、语义维度、项目和诗义的协调以及设计过程）。

这一项目有三个基本目标：通过对话为起点，促进当代设计和帕雷德斯家具公司间的创意合作；探究联合设计的过程，即由特邀设计师、专属使用者、企业及其资源构成的联合设计；按照设计制造出来的椅子进行拍卖，拍卖所得全部捐赠给联合国难民署。2012年"二重奏"项目设计出的椅子经由佳士得拍卖行拍卖，筹集到的11.5万欧元全部捐赠联合国难民署在非洲的教育和人道援助项目。

在构思、制作和策展方面，"二重奏"皆有侧重，分别体现在帕雷德斯本地企业、设计的诗意以及设计的社会责任感。

Duets
José Bártolo

Duets is a project with strong social responsibility and a humanitarian nature, that is focused on the many possibilities of designing, producing and using a chair, that involves concerns centred on the relationship between exclusive design and industrial production, and the diverse resulting cultural, economic and technology aspects (limited series; semantic dimensions; coordination between programme and poetry; design as a process).

The programme has three fundamental objectives: to facilitate dialogue as a starting point for creative stimulation between contemporary design and the furniture companies of Paredes; to explore the processes of co-design, involving an invited designer, an exclusive user, a business and its resources; to ensure for duet, the production of 5 contemporary and challenging chairs, whose auction revenues will go to the High Commissioner for Refugees. In the 2012 edition, the chairs produced within Duets were auctioned by Christie's, which turned over the amount earned (111,500 Euros) – to support UNHCR educational projects and humanitarian missions in Africa.

Thus, Duets positions itself at different levels in the contemplation, production and curating of design: from the contextual level of companies in the Paredes region, the poetic level of authorial creation, to the universal level of design's social responsibility.

葡萄牙总统阿尼巴尔·卡瓦科·席尔瓦 设计：保罗·罗柏 制作： ANTARTE	ANÍBAL CAVACO SILVA PAULO LOBO ANTARTE
企业家：贝纳通集团创始人卢西亚诺·贝纳通 设计： 路易斯·佩雷拉·米格尔 制作： CM CADEIRAS	LUCIANO BENETTON LUÍS PEREIRA MIGUEL CM CADEIRAS
建筑师；爱德华多·索托·德莫拉 设计： 设计工厂 制作： CUNHA MOBILIÁRIO	EDUARDO SOUTO DE MOURA DESIGN FACTORY* CUNHA MOBILIÁRIO
摄影师：扎尔梅 设计： 匿名建筑事务所、艾格尼斯·阿尔维斯 制作： MARGEM IDEAL	ZALMAÏ ARQUITETOS ANÓNIMOS MARGEM IDEAL
足球教练：何塞·穆里尼奥 因迪设计工作室 制作： ANTARTE	JOSÉ MOURINHO INDI ANTARTE
足球运动员：克里斯蒂亚诺·罗纳尔多 设计：妮妮·安德拉德·席尔瓦 制作： DISARTE MÓVEIS	CRISTIANO RONALDO NINI ANDRADE SILVA DISARTE
政治家：何塞·曼努埃尔·巴罗佐 设计： 保罗·帕拉 制作： JMS	JOSÉ MANUEL DURÃO BARROSO PAULO PARRA JMS
政治家：卢拉·达·席尔瓦 设计： 托尼·格里洛 制作： CM CADEIRAS	LULA DA SILVA TONI GRILO CM CADEIRAS

工作狂总统之椅

葡萄牙总统阿尼巴尔·卡瓦科·席尔瓦

设计：保罗·罗柏

制作：Antarte

2012

WORK PRESIDENT CHAIR

DESIGNED BY PAULO LOBO

PRODUCED BY ANTARTE

FOR ANÍBAL CAVACO SILVA

2012

工作狂总统之椅
设计：保罗·罗柏

Work President
Paulo Lobo

工作狂总统之椅

设计：保罗·罗柏

Work President

Paulo Lobo

阿尼巴尔·卡瓦科·席尔瓦于2006年当选葡萄牙总统，2011年1月23日连选连任。他是当代葡萄牙最重要的政治家之一，曾于1985年至1995年担任葡萄牙总理。

负责按照阿尼巴尔·卡瓦科·席尔瓦总统的性格设计椅子的是设计师保罗·罗柏，他在20世纪80年代创作的室内设计作品为他赢得了良好声誉。负责按照设计制造椅子的是Antarte，这是一家拥有强大技术能力、生产灵活性和高超手工艺水平的企业。

这把工作者风格的椅子含蓄地表达出它需要一张摆放井然有序的工作桌和足够的工作空间与之匹配。

这把由保罗·罗柏设计的名为"工作狂总统"的椅子，采用坚毅的线条配以低调的绿色皮革，它向我们讲述了一个哲理：在一把椅子形式的议程之下包含着生存的议程。

Aníbal Cavaco Silva was elected President of the Portuguese Republic in 2006, and re-elected on 23 January 2011. He is a politician with one of the most significant political careers in contemporary Portugal, having previously held the position of Prime Minister of Portugal between 1985 and 1995.

The designer **Paulo Lobo** was selected to design this chair, a designer with a strong reputation earned through interior design projects developed during the late 1980s.

The company responsible for production was **Antarte**, a company distinguished by its technological capacity and production flexibility, combined with the mastery of its craftsmen.

This chair, in *"Worker" style*, calls out for all that is missing to set the scene in the world that, implicitly, it should inhabit: an organized work desk with ample space for working.

葡萄牙总统阿尼巴尔·卡瓦科·席尔瓦
设计：保罗·罗柏
制作：Antarte

Work President chair
designed by **Paulo Lobo**
produced by **Antarte**
for **Aníbal Cavaco Silva**
2012

贝纳通之椅

企业家：贝纳通集团创始人卢西亚诺·贝纳通

设计：路易斯·佩雷拉·米格尔

制作：CM Cadeiras

2012

UNITED COLORS FOR BENETTON CHAIR

DESIGNED BY **LUÍS PEREIRA MIGUEL** IN COLLABORATION WITH **FILIPA OSÓRIO**

PRODUCED BY **CM CADEIRAS**

FOR **LUCIANO BENETTON**

2012

贝纳通之椅
设计：路易斯·佩雷拉·米格尔

United Colors for Benetton
Luis Pereira Miguel

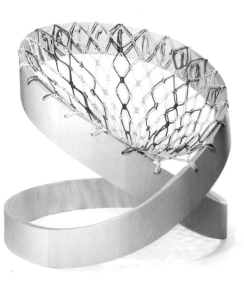

贝纳通之椅

设计：路易斯·佩雷拉·米格尔

United Colors for Benetton

Luis Pereira Miguel

卢西亚诺·贝纳通出生于1935年，是贝纳通集团创始人，热衷于服装和纺织品的市场和营销。贝纳通集团创办于1965年，现已扩展到全球120多个国家。建筑师和设计师路易斯·佩雷拉·米格尔受邀与菲利帕·奥索瑞欧一起为卢西亚诺·贝纳通设计一把椅子。路易斯·佩雷拉·米格尔于2005年创办自己的工作室，后于2008年赢得2008年国际色彩设计大赛冠军。

负责椅子制造的CM Cadeiras公司是一家具有较强技术优势的本土家具企业。

所设计出来的椅子体现了形式和技术的惊艳结合。其设计似乎完全背离人体工程学的规则，设计中巧妙融入了贝纳通品牌的特征，通过材料、语义和诗意价值的结合，实现对品牌和创始人的形象的表达。作品兼顾功能和美感，极具时代感。

Luciano Benetton, born in Treviso in 1935, is the founding entrepreneur of the Benetton Group, dedicated to the production and marketing of fashion and textiles, created in 1965 and currently located in 120 countries.

The architect and designer **Luís Pereira Miguel**, in collaboration with Filipa Osório was invited to design a chair for Luciano Benetton. Luís Pereira Miguel established his own atelier in 2005, after distinguishing himself by winning the international competition ColorsDesigner in 2008, promoted by POLI.Design and Benetton.

The company responsible for production was **CM Cadeiras**, a local reference in the production of furniture of great technical and formal complexity.

The chair resulting from this duet is an amazing formal and technical exercise. Its design seems to defy ergonomic rules, insinuating the Benetton brand, and incorporating materials, and semantic and symbolic values that reflect the brand and its founder. The resulting chair is contemporary, combining functionality and aesthetic value.

企业家：贝纳通集团创始人卢西亚诺·贝纳通
设计：路易斯·佩雷拉·米格尔
制作：CM Cadeiras

United Colors fot Benetton chair designed by **Luís Pereira Miguel** in collaboration with **Filipa Osório** produced by **CM Cadeiras** for **Luciano Benetton**
2012

（非）著名椅子
建筑师：爱德华多·索托·德莫拉
设计：设计工厂
制作：Cunha Mobiliário
2012

(UN)NOTED CHAIR
DESIGNED BY DESIGN FACTORY*
PRODUCED BY CUNHA MOBILIÁRIO
FOR EDUARDO SOUTO DE MOURA
2012

（非）著名椅子
设计：设计工厂

(Un) Noted
Design Factory*

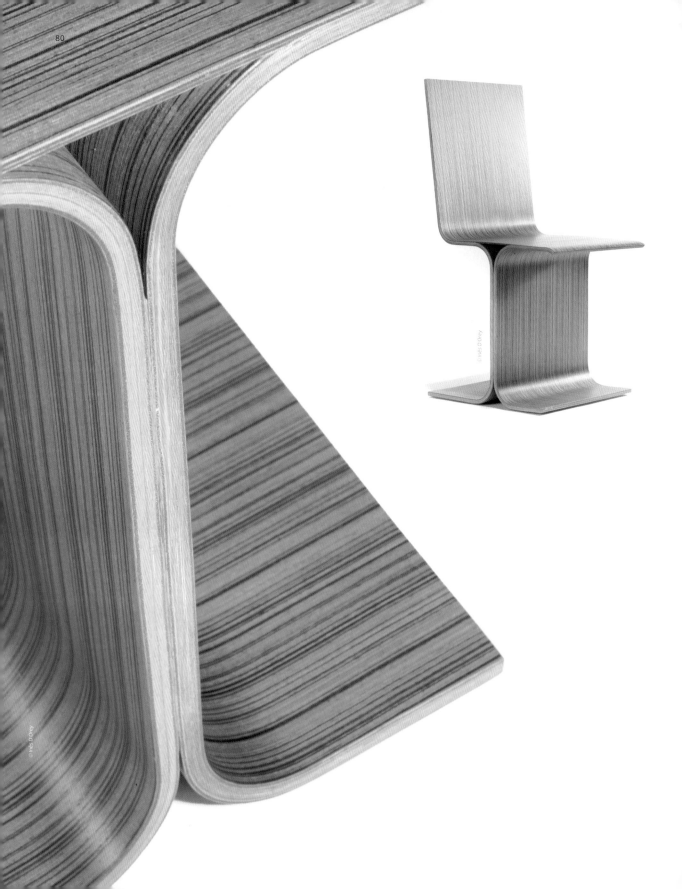

(非)著名椅子

设计：设计工厂

爱德华多·索托·德莫拉于1952年生于波尔图，是葡萄牙最有代表性的当代建筑师之一。他也是波尔图流派的代表性人物之一，费尔南多·塔乌拉和阿尔瓦多·西扎也是该流派重要代表。索托·德莫拉作为2011年普利兹克建筑奖的获得者是世界上最重要的仍在工作的建筑师之一。

设计工厂工作室受邀为爱德华多·索托·德莫拉设计椅子。设计工厂工作室成立于2007年，是RAR集团旗下的RAR Imobiliária公司的研究和设计工作室。负责按照设计制作椅子的是Cunha Mobiliário公司。Cunha Mobiliário拥有专业的生产和工艺技术，善于应对各类复杂挑战。

密斯·凡·德·罗的设计理念对索托·德莫拉有着深厚的影响，注重寻求形式上获得灵感和设计方案。设计中对"I"的轮廓形进行简约诠释，使用胶合板代替更常用的现代材料，这个设计既唤起人们对过去的遐想，又充满强烈的时代感，体现了索托·德莫拉别具一格的设计风格。

Cunha Mobiliário公司展示了其在执行和解决技术和构造细节难题方面的能力和严谨，采用两块有弧度的白杨木胶合板打造了最小版本的椅子。

建筑师：爱德华多·索托·德莫拉
设计：设计工厂
制作：Cunha Mobiliário

(Un) Noted

Design Factory*

Born in Porto, Portugal, in 1952, **Eduardo Souto de Moura** is one of the most representative contemporary Portuguese architects, a leading figure of the so-called Porto School, which includes Fernando Távora and Álvaro Siza Vieira. Awarded a Pritzer Architecture Prize in 2011, Souto de Moura is one of the world's most important working architects.

The atelier **Design Factory*** was invited to design the chair for Eduardo Souto de Moura. This atelier, established in 2007, is a research and design studio for RAR Imobiliária, a company belonging to the RAR Group.

The company responsible for the production of the **(Un)Noted** chair for the architect was the company **Cunha Mobiliário**. Cunha Mobiliário stands out for the quality of its production technicians and craftsmen capable of responding to complex challenges.

Recognizing Mies Van Der Rohe's influence in the work of Souto de Moura, the Design Factory, a design team for the Portuguese company RAR Imobiliário, sought formal inspirations and design solutions. The simplicity of the "I" profile is reinterpreted and the usual Modern materials are replaced by plywood, resulting in a chair that evokes historical references and affirms its contemporaneity. "(Un)Noted" emerges from this duet and calls to mind the distinctive character of the work of Souto de Moura.

In this project, Cunha Mobiliário showed its quality and rigour in implementing and solving technical and constructional details, thus enabling the Design Factory* to achieve the result of a minimal chair, from two curved surfaces of poplar-based veneered plywood.

(Un)noted chair designed by **Design Factory***
produced by **Cunha Mobiliário**
for **Eduardo Souto de Moura**
2012

灵魂之椅

摄影师：扎尔梅

设计：匿名建筑事务所、
艾格尼斯·阿尔维斯

制作：Margem Ideal

2012

Z'ALMA CHAIR

DESIGNED BY ARQUITETOS ANÓNIMOS IN
COLLABORATION WITH INÊS GONÇALVES

PRODUCED BY MARGEM IDEAL

FOR ZALMAÏ

2012

灵魂之椅
设计：匿名建筑事务所、艾格尼斯·阿尔梅斯

Z'alma
Arquitetos Anónimos

灵魂之椅

设计：匿名建筑事务所、艾格尼斯·阿尔维斯

扎尔梅出生于阿富汗喀布尔，是一位非常活跃的知名新闻摄影记者。1980年之后离开阿富汗，此后旅居瑞士洛桑。他曾在多个国际组织和非政府组织工作过，例如人权观察、国际红十字会、联合国难民署。

受邀设计椅子的是匿名建筑事务所，它践行着理念是没有人能独占自己的观点。"灵魂之椅"（Z'Alma）对于摄影师来说是一件很有"戏剧性"的事物。Margem Ideal公司负责将匿名建筑事务所和艾格尼斯·阿尔维斯的设计制作出来。

这把椅子的名字是"灵魂之椅"（Z'Alma），它用葡萄牙语发音会让人想起摄影师扎尔梅（Zalmaï）和葡萄牙语单词"Alma"，（表示灵魂）。灵魂是无形的，但是设计师希望这把椅子可以给他一个机会好好解读他的流浪和多面的特点。因此，设计师设计了这把有着雕塑感的椅子，一个有多重用途和解读的诗意物件。

Margem Ideal公司成立于1973年，是归属于Móveis Reina名下的公司。今年，这家公司投资于新市场和创新项目，产品特点是小型产品系列、对技术和构造细节的专注。

这把椅子的选材使用了废弃的软木，并废弃的胶合板和橡胶软木进行叠加造型，并使用了Margem Ideal的数控机床进行木材切割。

摄影师：扎尔梅
设计：匿名建筑事务所、艾格尼斯·阿尔维斯
制作：Margem Ideal

Z'alma chair designed by **Arquitetos Anónimos**
in collaboration with **Inês Gonçalves**
produced by **Margem Ideal** for **Zalmaï**
2012

Z'alma

Arquitetos Anónimos

Zalmaï, born in Kabul, Afghanistan, is one of the leading active photojournalists. Forced to flee his country after the Soviet invasion in 1980, he travelled to Lausanne, Switzerland. He worked for several international organizations and NGOs, such as Human Rights Watch, International Red Cross and the UN Agency for Refugees.

The **Arquitetos Anónimos** atelier was asked to design this chair for the photographer. This workshop practices the theory that no one exclusively owns their ideas.

"**Z'Alma**" is the name of the chair conceived as a "dramatic" object for the photographer. The "Z'Alma" chair, when pronounced in Portuguese evokes the name of the photographer (Zalmaï) and the Portuguese word "Alma" (soul). The soul is, by definition, incorporeal but the chair designed for Zalmai seeks to formally translate essential aspects of his nomadic and multifaceted identity. The result is a chair-sculpture, a poetic object that is open to various uses and interpretations.

The company **Margem Ideal** was a partner in the development of chair designed by Arquitetos Anónimos and Inês Gonçalves. The **Margem Ideal** company, founded in 1973 under the name Móveis Reina, is today a company that invests in new markets and innovative projects. Its production is characterized by small production series and by the quality of its technical and construction details.

"Z'Alma" uses cork waste and is constructed by overlapping layers of plywood and Rubber Cork, cut with CNC by Margem Ideal.

卢济塔尼亚人之椅

足球教练：何塞·穆里尼奥

设计：因迪设计工作室

制作：Antarte

2012

Lusíadas chair

designed by INDI

produced by Antarte

for José Mourinho

2012

卢济塔尼亚人之椅
设计：因迪设计工作室

Lusíadas
INDI

卢济塔尼亚人之椅

设计：因迪设计工作室

Lusíadas

INDI

何塞·穆里尼奥从2000年开始担任足球教练，是世界知名的葡萄牙人之一。他个性强烈，有着非凡的领导力和个人魅力。2004年加入英国切尔西俱乐部之后获得了一个世人皆知的绰号"特殊的人"。因为在2010赛季出色的执教表现，他在2011年1月被国际足联评为年度最佳教练。

受邀为他设计椅子的是来自ESAD（马托西纽什艺术和设计大学）工业设计系的研究团队。

设计师何塞·路易斯·费雷拉和胡安·皮图·弗莱雷设计了这把"卢济塔尼亚人之椅"。"卢济塔尼亚人"既指古代葡萄牙人，也是象征着路易·瓦兹·德·卡莫斯创作于16世纪的史诗，讲述了15和16世纪葡萄牙人航海探索的传奇故事。

"卢济塔尼亚人之椅"集合了葡萄牙木材、帆布、皮革以及通常用于运动领域的碳纤维框架元素。设计师试图通过这种材料的融合体现，是各种事件、行动、互动、反应和决定造就了这个男人、这位教练、这位世界上最重要的葡萄牙人之一。

Antarte公司负责制作"卢济塔尼亚人之椅"。这家公司具有很强的生产适应能力，善于提供解决方案，是葡萄牙国内的一流企业之一。由于设计非常复杂，这把"卢济塔尼亚人之椅"几乎全部由自动程序制作而成。

Undisputedly, **José Mourinho**, football coach since 2000, is one of the most internationally known Portuguese personalities. Renowned for his strong personality, leadership and charisma, he became recognized worldwide by the nickname "The Special One" following his hiring in 2004, by the English football club Chelsea. In January 2011 he was named Best Coach in the World by FIFA, in relation to the 2010 season.

INDI, the research group in Industrial Design from ESAD – College of Art and Design in Matosinhos, was invited to design the chair for Jose Mourinho. Designers José Luís Ferreira and Rui Pedro Freire, with an intense and award-winning body of work in industrial design, developed the chair **"Lusíadas"**. The name "Lusíadas" refers both to the general designation of the Portuguese people and to the epic poem by Luís Vaz de Camões, written in the 16th century, which tells the epic tale of the Portuguese maritime discoveries of the 15th and 16th centuries.

The "Lusíadas" chair combines Portuguese wood, canvas and leather, and structural elements of carbon fibre, materials often used in the field of sport . The duet sought to reflect the encompassing fabric of events, actions, interactions, reactions and decisions that made the man, and the coach, one of the most significant Portuguese personalities in the world.

The company **Antarte** was responsible for the production of "Lusíadas". Antarte, distinguished by its adaptable production capacity and its ability to provide solutions, is one of the leading companies in the domestic market. Due to its formal complexity, the chair for José Mourinho, "Lusíadas", was made almost entirely using automated processes.

足球教练：何塞·穆里尼奥
设计：因迪设计工作室
制作：Antarte

Lusíadas chair designed by **INDI** produced
ESAD – College of Art and Design, Matosinhos
by **Antarte** for **José Mourinho**
2012

C罗之椅

足球运动员：克里斯蒂亚诺·罗纳尔多

设计：妮妮·安德拉德·席尔瓦

制作：Disarte Móveis

2012

CR7 CHAIR

DESIGNED BY **NINI ANDRADE SILVA**

PRODUCED BY **DISARTE MÓVEIS**

FOR **CRISTIANO RONALDO**

2012

C罗之椅
设计：妮妮·安德拉德·席尔瓦

C罗之椅

设计：妮妮·安德拉德·席尔瓦

CR7

Nini Andrade Silva

克里斯蒂亚诺·罗纳尔多曾获得过2008和2014年国际足联颁发的金球奖，是当今世界足坛最杰出的球星之一。他于1985年出生于葡萄牙马德拉首府丰沙尔，他在葡萄牙竞技俱乐部接受训练，后于2003年加入曼彻斯特联队。自2006年起，这位葡萄牙国家队队长开始效力于皇家马德里足球俱乐部。

和克里斯蒂亚诺·罗纳尔多一样，妮妮·安德拉德·席尔瓦也出生于马德拉。妮妮·安德拉德·席尔瓦于2000年创立妮妮·安德拉德·席尔瓦工作室。自此，她开始成为具有国际知名度的设计师，不断受到追捧并获奖。

妮妮·安德拉德·席尔瓦表示，"C罗之椅"正是为罗纳尔多这位全球最具标志性的运动员设计的。这款贵妃沙发设计精美，完全按照运动员的体型设计，同时由于罗纳尔多出生于海岛，设计师从海浪和海滩岩石所蕴含的力量中获取灵感。

负责制作这把椅子的是成立于1965年的Disarte Móveis S.A.公司，在木质家具制造领域有着优良的传统和丰富的经验。

Cristiano Ronaldo is the world's best footballer according to the title awarded to him by FIFA in their annual ceremony. Born in 1985 in Funchal, Madeira, Portugal, he completed his sports training at Sporting Clube de Portugal before joining Manchester United in 2003. Since 2006, the captain of the Portuguese national team has played for Real Madrid.

Born in Madeira, like Cristiano Ronaldo, the designer **Nini Andrade Silva** created the Atelier Nini Andrade Silva in 2000, through which she has become internationally known, sought after, and awarded prizes.

According to Nini Andrade Silva, the **CR7** chair was designed as a work of art in the image of one of the most iconic athletes in the world. It is a Chaise-Lounge of sophisticated design, anatomically shaped to the body of the athlete, inspired by the power of symbols like the ocean waves and rocks from the beaches of the island where the player was born.

The company that produced the "CR7" chair was **Disarte Móveis S.A.**, founded in 1965, with a great tradition and experience in the manufacture of wooden furniture.

足球运动员：克里斯蒂亚诺·罗纳尔多
设计：妮妮·安德拉德·席尔瓦
制作：Disarte Móveis

CR7 chair designed by **Nini Andrade Silva**
produced by **Disarte Móveis**
for **Cristiano Ronaldo**
2012

图书馆椅

政治家：何塞·曼努埃尔·巴罗佐

设计：保罗·帕拉

制作：JMS

2014

LIBRARY CHAIR

DESIGNED BY **PAULO PARRA**

PRODUCED BY **JMS**

FOR **DURÃO BARROSO**

2014

图书馆椅
设计：保罗·帕拉

Library Chair
Paulo Parra

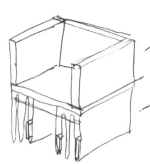

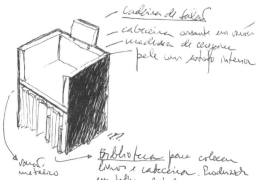

图书馆椅

设计：保罗·帕拉

Library Chair

Paulo Parra

何塞·曼努埃尔·巴罗佐是一名葡萄牙政治家和教师，于2004至2014年出任欧洲委员会主席，1985年任葡萄牙内政部副部长，1992年任外交部长，2002年至2004年任葡萄牙总理。他最小的儿子弗朗西斯柯是一名设计系学生。

保罗·帕拉是位设计学博士，现任FBAUL大学助理教授。他是一名专业设计师，也是一名研究员和收藏家。他同时还是MADE—葡萄牙埃武拉设计和艺术博物馆的负责人。

巴罗佐知识渊博，他说自己最喜欢的东西就是书。以此为灵感，保罗·帕拉设计了这把"图书馆椅"，它造型优雅，精致，选材和抛光的处理使得这款设计显出一种严肃的奢华。

JMS – J. Moreira da Silva & Filhos S.A.公司负责这把椅子的制作。JMS成立于1965年，是葡萄牙桌椅制造领域的龙头企业之一。

José Manuel Durão Barroso is a Portuguese politician and teacher, and ex-President of the European Commission, a position he held from November 2004 until 2014. In Portugal, he served as Under-Secretary of the Ministry of Internal Affairs (1985) and Minister of Foreign Affairs (1992). Between 2002 and 2004 he served as Prime Minister of the Portuguese Republic. His youngest son, Francisco, is a design student.

Paulo Parra has a doctorate in design and is currently Assistant Professor at FBAUL. He works professionally as a designer, and is also a researcher and a collector. He is the Director of MADE – Museu do Artesanato e do Design de Évora.

Possessing a vast cultural knowledge, Barroso confesses that his favourite object is the book. Starting from this inspiration, the Portuguese designer Paulo Parra designed a **Library Chair**, with a travel version in mind. The chair designed by Paulo Parra is elegant, sophisticated and soberly luxurious in the quality of its materials and finishes.

The production of this chair was given to the company **JMS – J. Moreira da Silva & Filhos**. Founded in 1965 it is one of the leading companies in Portugal in the manufacture of chairs and tables.

政治家：何塞·曼努埃尔·巴罗佐
设计：保罗·帕拉
制作：JMS

Library chair designed by **Paulo Parra**
produced by **JMS** for **Durão Barroso**
2014

光环之椅

政治家：卢拉·达·席尔瓦

设计：托尼·格里洛

制作：CM Cadeiras

2014

AURA CHAIR

DESIGNED BY **TONI GRILO**

PRODUCED BY **CM CADEIRAS**

FOR **LULA DA SILVA**

2014

光环之椅
设计：托尼·格里洛

Aura Chair
Toni Grilo

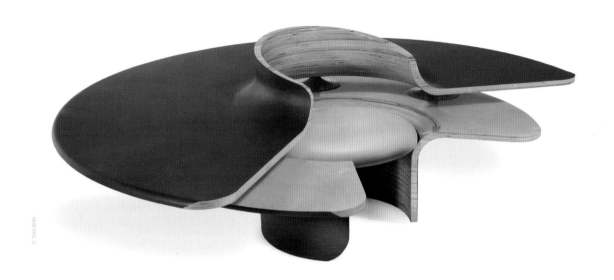
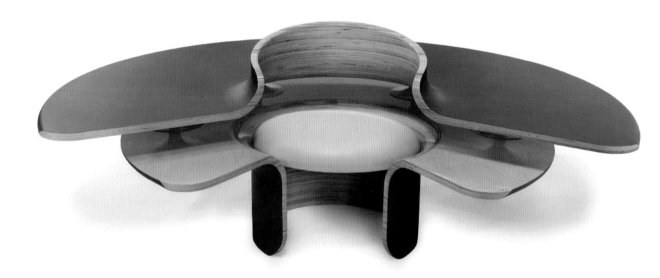

光环之椅

设计：托尼·格里洛

卢拉·达·席尔瓦是巴西第35任总统，任期自2003年1月至2011年1月，是巴西最受欢迎的政治家。他出生于伯南布哥，一个有着多元文化和深厚政治传统的地区。卢拉·达·席尔瓦极具个人魅力并且经历丰富，对当代拉美历史上的很多关键经济、社会和政治转型具有重要影响。

托尼·格里洛是受邀的设计师，他于1979年出生于法国，是一名法裔葡萄牙设计师和创意总监。他在葡萄牙找到了归属感，发现这是一个有着深厚工业和工艺基础的国度，并且爱上了这里的工艺技术和材料。通过在葡萄牙进行的几次合作，他决定永久定居在葡萄牙。2005年1月，他与设计师艾尔德·蒙泰罗联手成立了自己的第一家工作室Objection。2008他开始独立设计，在里斯本成立了自己的工作室，专注工业设计、家具和布景设计。

2013年托尼搬到了波尔图，担任了众多国际知名品牌的设计总监工作，包括Riluc，Haymann，Blackcork和具有百年历史的Topàzio。

这位巴西前总统表示，"设计一把椅子首先要考虑到工作的需要。我常常一坐就是好几个小时，我不能容忍一把带有休闲功能的椅子。可以说我从没坐着休息过。"这把由托尼·格里洛设计的"光环之椅"是椅子和桌子的结合，象征"工作的宝座"，采用坚硬橡木和包裹以金色皮革的坐垫构成。

负责制作"光环之椅"的是CM Cadeiras公司。这家公司也参与了2012年"二重奏"项目，其最具代表性的产品是为United Colors制作的贝纳通之椅。

政治家：卢拉·达·席尔瓦
设计：托尼·格里洛
制作：CM Cadeiras

Aura chair designed by **Toni Grilo** produced by **CM Cadeiras** for **Lula da Silva**
2014

Aura Chair

Toni Grilo

Considered the most popular politician in Brazil, and during his tenure, one of the most popular in the world, **Lula da Silva** was the 35th President of the Federative Republic of Brazil, a position he held from January 2003 to January 2011. Born in Pernanbuco, a state of diverse cultural richness and strong political tradition, Lula da Silva is a charismatic personality with an extraordinary personal experience, and a strong influence on some of the most important economic, social and political transformations in the contemporary history of Latin America.

Toni Grilo was the designer asked to participate in this unique duet. A Franco-Portuguese designer and creative director, Toni Grilo was born in Nancy (FR) in 1979. In Portugal, he found his roots and discovered a country with a rich industrial and artisanal heritage, and fell in love with the techniques and the material work involved. After several collaborations, he decided to settle permanently in Portugal: in January 2005, he founded his first agency, Objection, with the designer Elder Monteiro, but in 2008 he began his solo activities and opened his studio in Lisbon, specializing in industrial design, furniture and set design. In 2013, Toni moved to Porto and manages the creative direction of various international brands like Riluc, Haymann, Blackcork and the century-old Topázio.

According to the former President of Brazil, *"A chair should be designed for a work context. I spend many hours sitting, but I cannot associate a chair with leisure. Actually, I never sit down to rest"*. The chair "**Aura**", designed by Toni Grilo, is a hybrid between chair and desk, symbolically a "throne of work", produced in solid oak with a seat upholstered in gold leather.

The company **CM Cadeiras** was entrusted with the production of "Aura". This company, which was also involved in the 2012 edition of Duets, is recognized for its exemplary production of the United Colors for Benetton chair.

椅子游行

CHAIR PARADE

I HAVE NEVER FOUND IN THE COUNTRY SUCH A CONCERN IN CREATING A PIECE CONNECTED TO A MERELY FUNCTIONAL FACTOR (...) SINCE ART IS NORMALLY DISCONNECTED FROM FUNCTION, DISCONNECTED FROM PRACTICE, DISCONNECTED FROM HUMAN NATURAL ACTIVITY, WHICH IS THE CREATION OF UTENSILS.
NADIR AFONSO

我从没见过哪个国家会如此重视将设计与某一单纯功能因素结合的……因为艺术通常都是与功能、实用性和人类自然活动相分离的，艺术创作通常是区别于工具发明的。

——**亚冯梭**

椅子游行

桑德拉·劳

"椅子游行"项目始于2006年葡萄牙克里斯特罗小学。这个项目从瑞士设计师帕斯卡·纳普的"奶牛游行"处得到灵感,选取了日常生活中最常见、最普通但也最能代表这个地方的事物——椅子。它是家具中不可或缺的一部分,也是我们的生活态度、工作以及周围每个人现实世界的一部分。更为具体而言,它是家具工业不可或缺的一部分。此外,它还引起了人们对于椅子设计的关注度,激励了学生不断完善和设计更多椅子。椅子已经从普通物件变为专业设计作品。2014年"椅子游行"将有来自帕雷德斯10所中学和小学、21个家具制造商、约50位老师和超过900名学生参与。他们以葡萄牙设计史上著名的平面设计师、插画师和视觉艺术家为研究对象,进行研究、试验、制作、解构、重构、色彩鉴赏和材质分析等,总而言之,他们将根据不同的特征、理念、语言、传播和表达的含义对椅子进行重新打造。本次展览汇集了89把中的10把受20和21世纪葡萄牙所诞生的杰出的平面艺术作品所启迪而设计出的椅子设计作品。

葡萄牙著名抽象派艺术家那德尔·亚冯梭(1920-2013)曾于2007年在克里斯特罗小学说过:"我从没见过哪个国家会如此重视将设计与某一单纯功能因素结合的……因为艺术通常都是与功能、实用性和人类自然活动相分离的,艺术创作通常是区别于工具发明的。"

Chair Parade

Sandra Lau

Inspired by Cow Parade, by the Swiss Pascal Knapp, Chair Parade began in 2006 in Cristelo Elementary School. This project has taken the chair — daily object which symbolizes the furniture industry — as a basis. Trying to promote a taste for experimenting, Chair Parade has been raising students and school community's awareness of art and design. Furthermore it has been strengthening the bonds between the school community and the industry which employs more people in the Council. Chair Parade, as a recreational-pedagogical activity, has been contributing to the school absenteeism, failure and early drop-out reduction. In 2014 the coordination invited elementary and secondary students of all schools in Paredes (evolving about 50 teachers and more than 900 students) to modify chairs donated by 21 local manufacturers, getting inspiration from Portuguese graphic designers, illustrators and visual artists from 20th and 21st centuries. The work process included the study of a group of previously selected Portuguese graphic designers, illustrators and visual artists; the analysis of their languages, techniques and media; and finally, the design, the project, the (de)construction, in short, the transformation of common chairs into exclusive objects. This exhibition gathers 10 of the 89 chairs resultant of this interpretation exercise of the Portuguese graphic patrimony from the 20th and 21st centuries.

The Portuguese artist Nadir Afonso (1920–2013), during his visit to Cristelo Elementary School, highlighted the Chair Parade's importance and pioneer spirit: *"I have never found in the country such a concern in creating a piece connected to a merely functional factor (...) since art is normally disconnected from function, disconnected from practice, disconnected from human natural activity, which is the creation of utensils."*.

何塞·德·尼格雷鲁斯 帕雷德斯小学	**ALMADA-NEGREIROS** PAREDES BASIC SCHOOL
安德烈·达·罗巴 波尔塔小学	**ANDRÉ DA LOBA** BALTAR BASIC SCHOOL
海梅·马丁斯·巴罗塔 瑞博都萨小学和中学	**JAIME MARTINS BARATA** REBORDOSA BASIC AND SECONDARY SCHOOL
若·马查多 波尔塔中学	**JOÃO MACHADO** BALTAR SECONDARY SCHOOL
若·努涅斯 罗戴洛小学和中学	**JOÃO NUNES** LORDELO BASIC AND SECONDARY SCHOOL
路易·门德卡 卡萨玛依学院	**LUÍS MENDONÇA** CASA MÃE COLLEGE
R2设计 克里斯特罗小学	**R2 DESIGN** CRISTELO BASIC SCHOOL
胡安·门德卡 帕雷德斯中学	**RUI MENDONÇA** PAREDES SECONDARY SCHOOL
赛巴斯钦·罗德里格斯 索布雷拉小学和中学	**SEBASTIÃO RODRIGUES** SOBREIRA BASIC AND SECONDARY SCHOOL
赛巴斯钦·罗德里格斯 维莱拉小学	**SEBASTIÃO RODRIGUES** VILELA SECONDARY SCHOOL

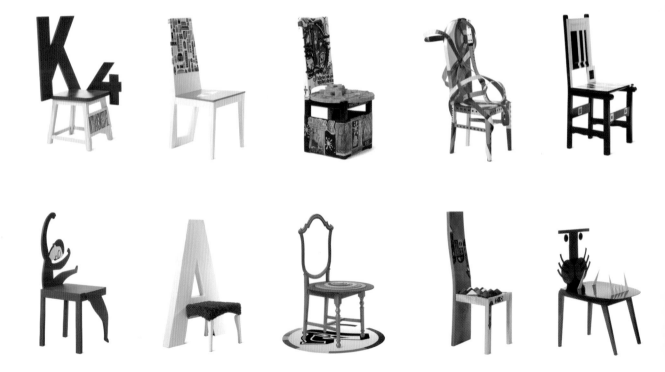

何塞·德·尼格雷鲁斯

帕雷德斯小学

学校
帕雷德斯小学
指导老师
R·伊莎贝尔·坎波斯
椅子捐赠
Cadeiras Machado公司
灵感
何塞·德·尼格雷鲁斯
尺寸
700 x 1140 x 560 cm

安德烈·达·罗巴

波尔塔小学

学校
波尔塔小学
指导老师
阿贝尔·达·席尔瓦
椅子捐赠
Fernando Dias da Silva & Filhos公司
灵感
安德烈·达·罗巴
尺寸
465 x 1025 x 500 cm

Almada-Negreiros

by Paredes Basic School

School
Paredes Basic School
Students from classes
9.º A and 9.º B
Teacher
R. Isabel Campos
Chair donated by
Cadeiras Machado
Inspired in
José de Almada-Negreiros
Dimensions
700 x 1140 x 560 cm

André da Loba

by Baltar Basic School

School
Baltar Basic School
Students from classes
7.º A, 7.º B, 7.º C and 7.º D
Teacher
Abel da Silva
Chair donated by
Fernando Dias da Silva & Filhos
Inspired in
André da Loba
Dimensions
465 x 1025 x 500 cm

海梅·马丁斯·巴罗塔
瑞博都萨小学和中学

学校
瑞博都萨小学和中学
指导老师
吉娜·查瓦斯
椅子捐赠
Zaga's公司
灵感
海梅·马丁斯·巴罗塔
尺寸
800 x 1000 x 800 cm

若·马查多
波尔塔中学

学校
波尔塔中学
指导老师
阿尔维诺·萨
椅子捐赠
CM Cadeiras公司
灵感
若·马查多
尺寸
500 x 1150 x 800 cm

Jaime Martins Barata
by Rebordosa Schools

School
Rebordosa Basic and Secondary School
Students from classes
9.º A
Teacher
Gina Chaves
Chair donated by
Zaga's
Inspired in
Jaime Martins Barata
Dimensions
800 x 1000 x 800 cm

João Machado
by Baltar Secondary School

School
Baltar Secondary School
Students from classes
11.º TD
Teacher
Avelino Sá
Chair donated by
CM Cadeiras
Inspired in
João Machado
Dimensions
500 x 1150 x 800 cm

若·努涅斯
罗戴洛小学和中学

学校
罗戴洛小学和中学
指导老师
孔塞桑·奥利维拉·路易斯·巴罗索
椅子捐赠
Custódio Machado公司
灵感
若·努涅斯
尺寸
540 x 1000 x 540 cm

路易·门德卡
卡萨玛依学院

学校
卡萨玛依学院
指导老师
费尔南达·库尼亚
椅子捐赠
Chairworld公司
灵感
路易·门德卡
尺寸
400 x 1000 x 400 cm

João Nunes
by Lordelo Schools

School
Lordelo Basic and Secondary School
Students from classes
C VOC 2
Teachers
Mª Conceição Oliveira and Luís Barroso
Chair donated by
Custódio Machado
Inspired in
João Nunes
Dimensions
540 x 1000 x 540 cm

Luís Mendonça
by Casa Mãe College

School
Casa Mãe College
Students from classes
9.º A
Teacher
Fernanda Cunha
Chair donated by
Chairworld
Inspired in
Luís Mendonça
Dimensions
400 x 1000 x 400 cm

R2设计
克里斯特罗小学

学校
克里斯特罗小学
指导老师
桑德拉·劳
椅子捐赠
Antarte**公司**
灵感
R2设计
尺寸
1400 x 1000 x 550 cm

胡安·门德卡
帕雷德斯中学

学校
帕雷德斯中学
指导老师
西莉亚·杜阿尔特
椅子捐赠
AFM Móveis**公司**
灵感
胡安·门德卡
尺寸
900 x 1060 x 900 cm

R2 Design
by Cristelo Basic School

School
Cristelo Basic School
Students from classes
5.º B, 6.º B and 6.º D
Teacher
Sandra Lau
Chair donated by
Antarte
Inspired in
R2 Design
Dimensions
1400 x 1000 x 550 cm

Rui Mendonça
by Paredes Secondary School

School
Paredes Secondary School
Students from classes
9.º B, 9.º E, 9.º J and 9.º I
Teacher
Célia Duarte
Chair donated by
AFM Móveis
Inspired in
Rui Mendonça
Dimensions
900 x 1060 x 900 cm

赛巴斯钦·罗德里格斯
索布雷拉小学和中学

学校
索布雷拉小学和中学
指导老师
埃尔莎·大卫
椅子捐赠
Cadeiras Miguel Machado 公司
灵感
赛巴斯蒂安·罗德里格斯
尺寸
440 x 1110 x 450 cm

赛巴斯钦·罗德里格斯
维莱拉小学

学校
维莱拉小学
指导老师
琼·法利亚
椅子捐赠
CM Cadeiras 公司
灵感
赛巴斯钦·罗德里格斯
尺寸
585 x 890 x 590 cm

Sebastião Rodrigues
by Sobreira Schools

School
Sobreira Basic and Secondary School
Students from classes
9.° D
Teacher
Elsa David
Chair donated by
Cadeiras Miguel Machado
Inspired in
Sebastião Rodrigues
Dimensions
440 x 1110 x 450 cm

Sebastião Rodrigues
by Vilela Secondary School

School
Vilela Secondary School
Students from classes
10.° VG
Teacher
Joana Faria
Chair donated by
CM Cadeiras
Inspired in
Sebastião Rodrigues
Dimensions
585 x 890 x 590 cm

葡萄牙帕雷德斯制造

Made by Paredes, Portugal

ABRITO S.A.
www.aabrito.com
geral@aabrito.com
Rua da Zona Industrial, 844 – Ap. 44
4584 – 908 Lordelo, Paredes
Tel.: (+351) 224 447 300
Fax: (+351) 224 447 308

ACASO
www.acaso.pt
cadeiracaso@gmail.com
Travessa da Abelheira
4580 – 607 Sobrosa, Paredes
Tel./Fax: (+351) 255 872 469

AFM MÓVEIS E MOBÍLIAS
www.moveisafm.com
Rua da Ferrugenta, 53
4580 – 473 Lordelo, Paredes
Tel.: (+351) 22 444 23 78
Fax: (+351) 22 444 08 58

ANTARTE
www.antarte.pt
antarte@antarte.pt
Avenida da Zona Industrial, 222A
4589–303 Rebordosa, Paredes
Tel.: (+351) 224 119 350
Fax: (+351) 224 119 351

ANTÓNIO SEABRA MÓVEIS
antonioseabramoveis@gmail.com
Rua das Castanheiras, 55 – Lote 6
Zona Industrial de Lordelo Ap. 101
4584 – 908 Lordelo, Paredes
Tel.: (+351) 914 211 633
Tel./Fax: (+351) 255 892 232

CADEIRAS LUÍS HENRIQUES
cadeirasluishenrique@iol.pt
Rua do Carvalhal, 100
4580–405 Duas Igrejas, Paredes
Tel.: (+351) 255 873 356

CADEIRAS MACHADO
cadeirasmachado@sapo.pt
Rua do Seixoso, 331
4580 – 372 Duas Igrejas, Paredes
Tel.: (+351) 224 097 276

CADEIRAS MIGUEL MACHADO
cadeirasmiguelmachado@gmail.com
Avenida Padre Manuel Pinto Preda, 887
4480 – 238 Duas Igrejas, Paredes
Tel.: (+351) 255 873 147 / 918 798 798

CADEIRAS TRISTÃO
cadeiras.tristao@gmail.com
Rua Alto do Facho, 200
4585–831 Rebordosa, Paredes
Tel.: (+351) 918 746 438

CADEIRAS VENÂNCIO E COELHO'S
www.cadeirasvenancio.com
cadeirasvenancio@gmail.com
Rua Cassil, 125
4585–112 Gandra, Paredes
Tel.: (+351) 224 110 176 / 916 189 342

CAMILA MÓVEIS
www.camilamoveis.pt
joseleal@camilamoveis.pt
Rua das Flores, 25
4585 – 424 Rebordosa, Paredes
Tel.: (+351) 224 442 565 / 964 553 232

CHAIRWORLD
www.chairworld.pt
geral@chairworld.pt
Rua do Fôjo Velho, 284
4585 – 425 Rebordosa, Paredes
Tel.: (+351) 224 156 983 / 224 110 785
Fax: (+351) 224 161 009

CASTRO SOUSA MÓVEIS
www.castrosousamoveis.com
castrosousamoveis@gmail.com
Rua Nova do Cerno, 120
4585–503 Rebordosa, Paredes
Tel./Fax: (+351) 224 112 751

CM CADEIRAS
www.cadeiras.net
geral@cadeiras.net
Rua Zona Industrial, 348
4589 – 907 Rebordosa, Paredes
Tel.: (+351) 224 156 509
Fax: (+351) 224 156 968

CÓMODOS MOBILIÁRIO
comodosmobiliario@gmail.com
Estrada Nacional 209, 5585
4580 – 439 Lordelo, Paredes
Tel.: (+351) 917 769 660

CUNHA MOBILIÁRIO
www.cunhamobiliario.pt.vu
Rua da Vila, 361
4580 – 472 Lordelo, Paredes
Tel.: (+351) 224 442 630

CUSTÓDIO CARMO MACHADO
ccm.mobiliario@sapo.pt
Rua Rui Barros, 41
4580– 410 Lordelo, Paredes
Tel.: (+351) 255 873 224

DISARTE
www.disarte.pt
disarte.home@gmail.com
R. Ilidio Ferreira 151
4584 – 908 Lordelo, Paredes
Tel.: (+351) 224 447 550 / 917 202 728

EVANYROUSE
www.evanyrouse.pt
geral@evanyrouse.pt
Rua Nova de Monte Alto, 40
4585 – 466 Rebordosa, Paredes
Tel.: (+351) 224 445 068
Fax: (+351) 224 151 149

FENABEL
fenabel@fenabel.com
www.fenabel.com
Rua das Fontaínhas, 162 – Ap. 32
4589–907 Rebordosa, Paredes
Tel.: (+351) 224 119 120
Fax: (+351) 224 119 129

FERNANDO DIAS DA SILVA & FILHOS
Rua Nossa Senhora da Ajuda, 51
4585–341 Rebordosa, Paredes
Tel.: (+351) 224 114 384
Fax: (+351) 224 114 384

FERNANDO LEITE & FILHOS
export@fleite.pt
Rua Nossa Senhora da Ajuda, 51
4585 - 341 Rebordosa, Paredes
Tel.: (+351) 224 151 589

GUALTORRES
www.gual.pt
gual@gual.pt
Rua da Zona Industrial, 1189
4580 - 565 Lordelo, Paredes
Tel.: (+351) 224 449 173
Fax: (+351) 224 449 175

GUARNIÇÃO
www.guarnicao.com
luis.barbosa@guarnicao.com
Rua da Fábrica, 272
4585 - 053 Baltar, Paredes
Tel.: (+351) 224 152 602
Fax: (+351) 224 152 604

IDEALMOVEL
www.idealmovel.pt
geral@idealmovel.pt
Rua Dr. José B Tavares, 78
4580 - 593 Paredes
Tel.: (+351) 255 788 410
Fax: (+351) 255 788 419

IDEAMOBILE
www.ideamobile.pt
info@ideamobile.pt
Rua Pinto de Araújo, 367/375 - 2.º direito
4450-777 Leça de Palmeira
Tel.: (+351) 229 967 976; 229 942 517/8
Fax: (+351) 229 967 977

JMS
www.jms.pt
info@jms.pt
Avenida das Fontainhas, 42 – Ap. 22
4589-907 Rebordosa, Paredes
Tel.: (+351) 224 157 390/320
Fax: (+351) 224 114 477

**JOSÉ FERNANDO LOUREIRO
DOS SANTOS**
www.jflscadeiras.com
geral@jflscadeiras.com
Zona Industrial de Rebordosa – Ap. 83
4589 - 907 Rebordosa, Paredes
Tel.: (+351) 224 113 581 / 224 152 637 /
922 020 594
Fax: (+351) 224 110 515

KUARTRUS INTERIORES
www.kuatrus-interiores.pt
kuatrus@gmail.com
Estrada Nacional 15
Rua Central de Gandra, 240
4585-116 Gandra, Paredes
Tel.: (+351) 916 113 086

LIGA MÓVEIS/ LIGA MELF
www.ligadomobiliario.com
silvestre.sc@ligadomobiliario.com
Zona Industrial de Lordelo – Ap.120
4584-908 Lordelo, Paredes
Tel.: (+351) 224 447 360 / 918 030 613

MARGEM IDEAL
www.margemideal.com
margem.geral@gmail.com
Rua de Parteira, 330
4580-548 Lordelo, Paredes
Tel.: (+351) 224 440 418
Fax: (+351) 224 440 431

MÓVEIS FIALHO
www.moveis-fialho.com
geral@moveis-fialho.com
Rua Quintã de Baixo, 117
4585-505 Rebordosa, Paredes
Tel.: (+351) 224 111 436 / 919 529 192

NOBILIS
www.nobilis.pt
geral@nobilis.pt
Rua Multipark I, 151
4595-542 Seroa, Paredes
Tel.: (+351) 255 891 061
Fax: (+351) 255 891 063

RABISCOS SENSATOS
facebook.com/Rabiscossensatosfurniture
rabiscossensatos@gmail.com
4580 Lordelo, Paredes
Tel.: (+351) 912 856 957

RONFE CLASSIC
www.ronfe.com
comercial@ronfe.com
Rua da Zona Industrial, 320
4580 Lordelo, Paredes
Tel.: (+351) 255 880 490
Fax: (+351) 255 880 499

SUIDENETO
www.suideneto.com
geral@suideneto.com
Rua de S. Miguel, 497
4585-175 Gandra, Paredes
Tel.: (+351) 224 159 777
Fax: (+351) 224 159 779

WOODSPACE
www.woodspace.pt
(1) Travessa Zona Industrial, 3/9
4585-303 Rebordosa, Paredes
Tel.: (+351) 224 110 787 / 961 577 259
Fax: (+351) 224 113 455
(2) Avenida D. Manuel II, 1362
4470-334 Maia
Tel.: (+351) 229 471 691 / 961 577 258
Fax: (+351) 229 471 693
(3) Alameda dos Oceanos, L4.4301.O Bloco
1990-118 Lisboa
Tel.: (+351) 961 577 256

ZAGAS - AEF MEUBLES
www.zagas.pt
comercial@zagas.pt
Rua Rainha Santa Teresa, 467 – Ap. 116
4589 - 907 Rebordosa, Paredes
Tel.: (+351) 255 880 590
Fax: (+351) 255 880 598

葡式设计2000–2014

HOW TO PRONOUNCE DESIGN IN PORTUGUESE? (2000 – 2014)

CURATOR

BÁRBARA COUTINHO
DIRECTOR OF MUDE – MUSEU DO DESIGN E DA MODA FRANCISCO CAPELO'S COLLECTION

策展人

芭芭拉·科蒂尼奥
里斯本艺术设计博物馆馆长

"BEING PORTUGUESE" HAS TODAY, KNOWING THAT AMONGST THE STRONGEST CHARACTERISTICS OF PORTUGUESENESS, WE FIND UNIVERSALITY, INTERCULTURALISM AND "THE INFINITE CAPACITY TO BE A MULTITUDE OF THINGS AT THE SAME TIME, EVEN IF THEY APPEAR CONTRADICTORY"
AGOSTINHO DA SILVA

作为葡萄牙人的身份在今天的意义、目的和有效性。尽管存在一些非常典型的葡萄牙主义的特征，我们还是发现了一些特征是普世的、跨文化的和同时代表多种事物的，即使存在相互矛盾的一面。

——奥古斯丁·达·席瓦尔

Brands

葡式设计2000-2014

芭芭拉·科蒂尼奥

"葡式设计"展览旨在阐释葡萄牙的地理位置、历史遗产、文化传统以及集体意识对于设计师创作的影响，探索葡萄牙设计的特征。展览的策划初衷是反思作为葡萄牙人的身份在今天的意义、目的和有效性。尽管存在一些非常典型的葡萄牙人特性，我们还是发现了一些特征是普世的、跨文化的和同时代表多种事物的，即使存在相互矛盾的一面。葡萄牙设计到底可以在多大程度上跳出我们的传统文化框架，从而为这个高效的多文化世界做出贡献呢？葡萄牙人精神可否通过设计体现出来呢？

本次展览旨在回答这样的一个问题。展览主要选取了2000年至2014年间的家具设计，这一时段的主要特征是部分葡萄牙设计师的国际化趋势，来自不同教育背景、社会关切和意识的新兴一代的到来，以及一些追求精益求精并保留独特技艺的传统制造行业的现代化发展。本展在设计之初就考虑到了旅行的目的，仓储和运输系统也同时变身展览构图和框架。这个展览共有四个主题，展示了不同时期设计师和建筑师的61件作品。

在"探寻事物本质"主题之下，展品强调简约、纯净，去除所有肤浅的元素，突出展品最简单的架构。"材料和实物的实用智慧和情感"从欣赏和实际调查出发，探寻各种材质的物理行为、美学和结构潜力，追求对物体类型的再创作。"大众与博识"在再造材料文化的过程中关注传统生活模式和模型，使用传统语言，并且发掘通常被遗忘的传统工艺领域的特殊知识。作品体现传统精巧技艺和现代大众文化的结合，对空间、归属和记忆提出质疑。在最后的"这不（仅仅）是个烟斗"主题下，我们会看到一些智慧、幽默而讽刺的设计，充满暗示、趣味、荒诞和暗喻。它们体现对材料文化的概念性调查，研究每个物体不同层次的含义，探索其诗意和传播价值，仿佛在展示一种"有形的诗歌"。

2014年11月，该展览还将在里斯本和帕雷德斯两个展示中心展出，以家具设计展示为主，同时包含部分传播和服饰设计展示。展览将按照年代顺序展开，追溯至1980年，此外还包含部分1950年至1970年间的葡萄牙传统设计作品和物件，以此回应最初的提问。

How to Pronounce Design in Portuguese? (2000 – 2014)

Bárbara Coutinho

The *How to pronounce design in Portuguese?* exhibition aims to understand how Portugal's geographical situation, heritage, culture, traditions and collective consciousness have influenced the work of each designer, and explore the identity of Portuguese design. The exhibition arises from reflection on the meaning, purpose and validity that the condition of *"being Portuguese"* has today, knowing that amongst the strongest characteristics of Portugueseness, we find universality, interculturalism and "the infinite capacity to be a multitude of things at the same time, even if they appear contradictory" (Agostinho da Silva). To what extent does the affirmation of Portuguese design not occur in the exact manner that these symbols of our cultural framework were developed, thereby contributing to an effectively global and multicultural world? Can the spirit of Portugueseness be translated into design?

This exhibition seeks to respond to that question. Centred on furniture designed between 2000 and 2014, a period marked by the consolidated internationalisation of certain Portuguese designers, by the arrival of a new generation with different training, social concerns and consciousness, and by the modernisation of certain traditional manufacturing sectors that focused on excellence, human resources and the unique technical know-how that still exists in Portugal. Designed from the outset for the purpose of travelling, the system of storage and transport simultaneously acts as its graphics and exhibition framework. Organised in four main nuclei, the exhibition presents 61 pieces by designers and architects from various generations.

In the nucleus **Search for the Essence of Things**, the pieces share the same asceticism and brevity, suggesting a simplicity and purity that seeks to eliminate any superficial element, highlighting the minimal architecture of each object; in the nucleus **Practical Intelligence and Sensibility for the Material and Concrete**, proposals arise from an appreciation and practical investigation, very often experimental, of the physical behaviour and potential (aesthetic and structural) of materials and their transformation, seeking to reformulate the different object typologies; in the nucleus **Between the Popular and the Erudite**, material culture is reinvented by looking at the models and modes of traditional life, working with traditional languages and recovering specific knowledge of craftsmanship (often, almost forgotten). This work can acquire a certain erudite quality and be representative of our secular cultural blend, and question the concepts of place, belonging and memory; and finally, in the nucleus **This is not (just) a Pipe**, we are confronted with objects that are sustained by intelligent humour and irony, and which value suggestiveness, playfulness, nonsense, quotation and metaphor. They represent a conceptual investigation into material culture and work on the possible levels of meaning of each object, exploring its poetics and communicational value, almost like "concrete poetry". In November 2014, this exhibition will be held in two complementary and simultaneous centres (Lisbon/MUDE and Paredes) focused on furniture design with some incursions into communication design and fashion design. In chronological terms, it will stretch back to 1980, and include some traditional Portuguese pieces and objects designed between 1950-70, in response to the initial question.

圣地亚哥

阿尔瓦罗·西扎

Santiago Stool

Álvaro Siza Vieira

这件圣地亚哥的座椅以其纯粹的质感，精致的材料，及其体现的严密理性，代表了设计师阿尔瓦罗·西扎的审美。这件作品的设计从某种程度上延续了传统葡萄牙座椅的风格。它去繁从简，毫无造作之感。外形和功能仅保留了最简要的部分。

The Santiago stool demonstrates the rational purity, the excellence of materials and the rationalist rigour that characterises the aesthetic of Álvaro Siza Vieira. Bringing a certain familiarity from traditional Portuguese tools, it is a visible exercise in reducing form and function to the essential, without any artificiality or unnecessary elements.

阿尔瓦罗·西扎
凳子
老苏格兰松树和桃花心木
30x30x42厘米
制造：SPSS Lda
1993

Álvaro Siza Vieira
Santiago Stool
Old scots-pine wood, and mahogany wood
30x30x42 cm
Production by SPSS Lda
1993

1. 探寻事物本质 / *Search for the Essence of Things*

平地

路易斯·基斯塔斯

Flatland seats A and B

Luis Giestas

该设计运用简单平面的交叉，使椅子的功能性、稳定性和力量感得以保证。看似尚未完成的设计感，椅子拆装容易，造型极其简约独特。

Function, stability and strength are guaranteed through the intersection of simple planes. Unfinished and easily disassembled, these stools are unique for their extreme simplicity.

路易斯·基斯塔斯
两个可拆卸的座椅
桦木胶合板
A：32x48.5x53厘米 | B：51.5x46x72.5厘米
原型
2010

Luis Giestas
Flatland seats A and B
Two small demountable seats
Birch plywood
A: 32x48.5x53 cm | B: 51.5x46x72.5 cm
Prototype
2010

穆别里多什

Pedrita（丽塔·若和佩德罗·费雷拉）

Mobiletos

Studio Pedrita

这件作品作为"穆别里多什"家族的一员，作用介于装饰配件与家具之间。它的设计遵从日常家庭生活习惯，考虑到了箱子的不同型号，有不同的高度和方向。该作品显示了与我们的生活空间及使用物品之间的关系。

Part of the "Mobiletos" family, this piece is between accessory and furniture. The result of observing the habits and practices of household living, various models of boxes, with differing heights and directions, suggest other relationships with the space we inhabit and the objects we use.

Pedrita（丽塔·若和佩德罗·费雷拉）
多用途家具
原型#2
天然桉木、无光清漆和半光漆
73x118x20厘米
限量版
出品： Show me Design & Art艺廊
2011

Studio Pedrita (Rita João and Pedro Ferreira)
Mobiletos – Mobileto prototype #2
Multipurpose piece of furniture
Natural eucalyptus wood
73x118x20 cm
Limited Edition by Show me Design & Art Gallery
2011

© Manuel Chicharro

衣服

安娜·瑞尔瓦

Clothing chair

Ana Relvão

我们很清楚在哪放脏衣服，在哪放洗好的衣服。可是在哪放准备再穿的衣服呢？这件展品给出了答案。它类似一件简易实用版本的床头柜。

We know where to put dirty clothes, we know where to put washed clothes, but where do we put clothes that have been used but which we want to use again? The answer is provided by a simple, elementary and practical form, in a contemporary reinterpretation of the nightstand.

安娜·瑞尔瓦
衣挂
回收木材
45x75x 27.5厘米
制造／出品：Remix Project
2013

Ana Relvão
Clothing chair
Wood waste
45x75x 27.5 cm
Production/Edition by Remix Project
2013

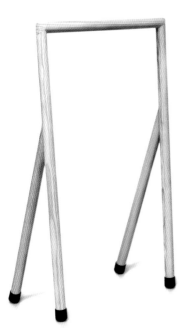

WR.02
马可·索萨·桑托斯

WR.02 – Skin Chair
Marco Sousa Santos

设计师马可·索萨·桑托斯从美学及结构角度研究了每种材料，及可对各种材料进行改革的技术资源。这件作品是他的实验成果之一。它结合了传统工序和现代技术，我们可以把WR.02看作对椅子原型的体现。这件作品以其轻盈，稳固，精致的抛光及纯粹的设计著称。

This chair is an example of the experimental research that Marco Sousa Santos has undertaken on the aesthetic and structural qualities of each material and the technological resources available for its transformation. Associating traditional processes with new technologies, WR.02 can be read as a reflection on the chair archetype and is distinguished by its extreme lightness, strength and stability, sophisticated finishes and aesthetic purity.

马可·索萨·桑托斯
椅子
橡胶涂层实心榉木
43x47x78厘米
制造：Branca-Lisboa
2009

Marco Sousa Santos
WR.02 – Skin Chair
Solid beech wood with rubber coating
43x47x78cm
Production by Branca-Lisboa
2009

菲戈	"Figo" Carpet
米格尔·维埃拉·巴普蒂斯塔	Miguel Vieira Baptista

该设计由相同大小的彩色六边几何图形构成。米格尔·维埃拉·巴普蒂斯塔大胆引入了空间设计，有意打破了原始图案，但与整体结构毫无冲突之感。

A geometric and coloured set of identically sized hexagons. Miguel Vieira Baptista introduces unexpected empty spaces that deliberately interrupt the initially established pattern, but without contradicting the formal structure.

米格尔·维埃拉·巴普蒂斯塔
地毯
羊毛
300x190厘米
限量版1/40
出品：Loja da Atalaia
2003

Miguel Vieira Baptista
"Figo" Carpet
Wool
300x190 cm
Ed. 1/40
Limited Edition by Loja da Atalaia
2003

墙毡

菲利普·阿拉尔孔

Wall Felt

Filipe Alarcão

这件墙面毛毡的设计灵感源自一个从透视角度观察的立方体。它探索了平面与立体的关系。毛毡的颜色构成给人以颜色深浅渐进的奇特感觉。由羊毛及软木橡树皮制成的毛毡，隔音保暖，又赋予了整体空间舒适感。这件作品集功能性与装饰性于一体。

This wall felt arises from the depiction of a cube in perspective and explores the relationship between planes and volumes. The colour composition suggests an illusion of depth, while at the same time the wool and cork felt confers comfort to the space in which it is used, due to its qualities of thermal and acoustic insulation. It is situated between the functional and the decorative.

菲利普·阿拉尔孔
墙毡
羊毛毡，软木
100x120x3厘米
Hand Matters出品
2011

Filipe Alarcão
Wall Felt
Wool felt, cork
100x120x3 cm
Edition by Hand Matters
2011

布顿

琼娜·卡瓦略

Buton chair

Joana Carvalho

这件作品的外形设计古典。"布顿"可完全被微生物降解,因此对环境无污染。另外,出于对环境保护和可持续发展的考虑,所有材料,包括自然材料和来自纺织工厂的废弃材料,都经过精心的选择。同样重要的是,这件作品的设计应用了传统制作工艺。

Classic in its form, Buton is totally biodegradable and non-polluting. Arising from a concern with questions of sustainability and respect for the environment, all the components were chosen with care and rigour, whether they are natural materials or waste from the textile industry. Equally important is the use and appreciation of traditional manufacturing techniques.

琼娜·卡瓦略
椅子
实心榉木,自然色清漆或涂漆,顶部和背面采用了海洋纤维材料
350x50x75厘米
出品:Margem Ideal
2012椅子上的艺术参展作品

Joana Carvalho
Buton chair
Solid beech wood, finishing in natural colour varnish or lacquered, top and back in maritime fibre wood, Burel seat
350x50x75 cm
Edition by Margem Ideal
Presented under the project Art on Chairs 2012

折痕
佩德罗 · 席尔瓦 · 迪亚斯

Crease Chair
Pedro Silva Dias

构思与艺术：一片铝箔折叠成四部分。该作品外形设计简单精巧，折叠的铝箔完全构成了椅子的形状，有座椅和靠背。同时，折叠的座椅部分形成了微妙的投影。这正体现了设计师佩德罗 · 席尔瓦 · 迪亚斯作品的特点：看似枯燥，又极富巧思的形式主义。

Ingenuity and art: A sheet of aluminium folded in four. With a simple, minimal and rigorous form, this gesture is the essence of the chair. The folded and cut sheet forms a chair, with its seat and back. At the same time, the fold also creates a subtle play of reflections and shadows, and is an example of the dry and intelligent formalism that characterises the work of Pedro Silva Dias.

佩德罗 · 席尔瓦 · 迪亚斯
椅子
铝
75x45x37厘米
原型
2000

Pedro Silva Dias
Crease Chair
Aluminum
75x45x37cm
Prototype
2000

D&D II

何塞·维亚纳

D&D chair II

José Viana

该作品取自何塞·维亚纳关于可折叠成平面的椅子的实验。它基于20世纪80年代末举办的波尔图设计上的一项提议，独特之处在于用材极省，且节约占地空间。这件椅子设计简单，方便携带存放，代表了微型建筑的实用性。

An exemplary piece from the research that José Viana has been carrying out on chairs that can fold flat. Following a proposal produced for Proto Design at the end of the 1980s, this piece is remarkable for its economy of materials and reduced occupation of space. Simple, easy to carry and store, this chair is also representative of a minimal architecture that ensures its usefulness.

何塞·维亚纳
椅子
一对水平相同的椅子
桃花心木和榉木胶合板1.2厘米和0.9厘米
87x37x37厘米（打开）
限量版
2004

José Viana
D&D chair II
Pair of same-level chairs
Mahogany and beech plywood measuring 1.2cm and 0.9mm
87x37x37 cm (open)
Limited Edition by José Viana
2004

2. 材料和实物的实用智慧和情感
Practical Intelligence and Sensibility for the Material and Concrete

W.01 壳

马可·索萨·桑托斯

W.01 – Shell Chair

Marco Sousa Santos

这件作品是Branca Lisboa家具和照明系列的设计作品之一，展现了对于细节的追求、精巧的美感和雕塑的气质。通过贝壳造型、木材、新兴制造技术和实验性设计品位实现了传统和现代的完美结合。

One of the pieces that make up the Branca Lisboa collection of furniture and lighting by Marco Sousa Santos. The perfection of detail, the refined aesthetic sense and a particular sculptural quality characterise this chair which combines tradition (the conch shape and wood) and modernity (new manufacturing technologies and a taste for experimentalism).

马可·索萨·桑托斯
椅子
桦木胶合板
70x78x70厘米
制造：Branca-Lisboa
2009

Marco Sousa Santos
W.01 – Shell Chair
Birch plywood
70x78x70cm
Production by Branca-Lisboa
2009

Hk 0.75
MO-OW 设计工作室

Hk 0.75 sideboard
MO-OW design

简约的直线设计使得这款餐具柜显得与众不同。设计师从建筑学获得灵感,将每件家具设计成一个宜居空间。采用传统技艺精致加工木料,体现设计师对材质的鉴赏和工艺的追求。

Straight, simple and minimalist lines distinguish this sideboard. Ângela Frias and Gonçalo Dias use their architecture as a starting point to design every piece of furniture as a habitable space. Built using the finest traditional wood working techniques, they reveal a special appreciation for the material and an awareness of the importance of the quality of execution.

MO-OW 设计工作室
餐具柜
梧桐木和涂漆中密度纤维板
160x50x85厘米
制造:MO-OW设计
2012

MO-OW design (Ângela Frias and Gonçalo Dias)
Hk 0.75 sideboard
Sycamore wood and lacquered MDF
160x50x85cm
Production by MO-OW design
2012

宝石

菲利普·阿拉尔孔

Gem square table

Filipe Alarcão

这款桌子突出纯色的整体、毫无手工制造的痕迹。这正体现设计师对于纯粹主义和几何美学的追求，也体现设计师在一件作品中对不同材质和精巧技术的融合。

Monochromatic and apparently monolithic, without traces of being worked by hand, the Gem table is representative of the purist and geometric poetics of Filipe Alarcão, as well as being a good example of the different materials and the formal and technical refinement that he seeks in each piece.

菲利普·阿拉尔孔
方桌
里斯本系列
可丽耐
26x75x75厘米
制造：Tema Home
2009

Filipe Alarcão
Gem square table
Lisbon Collection
Corian
26x75x75cm
Production by Tema Home
2009

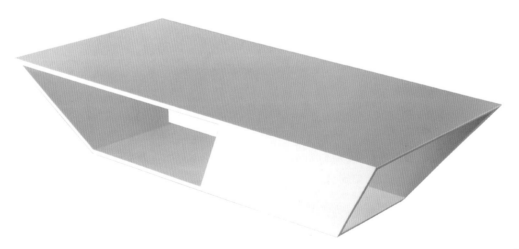

D-透视

米格尔·维埃拉·巴普蒂斯塔

D-Perspective tables

Miguel Vieira Baptista

D-透视桌子系列带给参观者一种出乎意料和不确定性。尽管线条平直、简约大方,但不同的高度和倾斜的桌腿制造出一种微妙的破坏性,既吸引眼球又引人思索。观察者从不同角度会得出不同的解读。在古典主义中加入的出人意料和"不和谐音"正是设计师创作的特征之一。

The D-perspective tables create surprise and uncertainty in the observer. Behind their rational rigour and minimal simplicity, the different height and inclination of the legs creates a subtle subversion that catches the eye and leads one to question. Depending on your point of view, the reading of the piece alters. The introduction of the unexpected or a "dissonant note" that counteracts the initial classicism, is one of Miguel Vieira Baptista's characteristics.

米格尔·维埃拉·巴普蒂斯塔
桌子
桦木和桦木胶合板
120x49x33厘米 | 45x45x 33厘米
原型
2009

Miguel Vieira Baptista
D-Perspective tables
Birch and birch plywood
120x49x33cm | 45x45x 33cm
Prototype
2009

Puf Fup
安娜·梅斯特

Puf Fup
Ana Mestre

采用2500颗自然完美的软木珠制成的链子。软木光滑的表面、轻便柔滑的质感使人感到非常舒适。使用者可以充分发挥创意进行使用。

A long chain of 2,500 perfect, natural cork spheres, each one 2.5 cm in diameter. The smoothness, texture and lightness of the cork ensure comfort, a soft touch and a pleasing relationship with the human body. It can be sat on in a variety of ways, according to the disposition and number of coils of its spheres, a possibility that challenges the creativity of its user.

安娜·梅斯特
软木球
65x65x35厘米
出品：Corquedesign.com
2005

Ana Mestre
Puf Fup
Cork spheres
65x65x35cm
Edition by Corquedesign.com
2005

CUT furniture chair
Mariana Costa e Silva

"切割"意味着尽可能少地使用材料、劳动力、工具和空间。托架、椅子、桌子和扶手椅皆由合成木板制成，不需要胶水、工具或者螺丝。选取这一材料主要考虑其受力性强、生态环保、防水性好和色泽一致性的特性。特殊卡口的设计非常方便使用者进行拆装。

Cut furniture is synonymous with less material, less labour, less tools, less space. Stools, chairs, tables and armchairs emerge from sheets of agglomerated wood fibre without needing a single screw, tool or glue. This material was chosen for its mechanical strength and ecological properties, its water resistance, and for its uniform colour, which reduces the need for finishes. It is quickly and easily assembled and disassembled by the user through a system of interlocking parts.

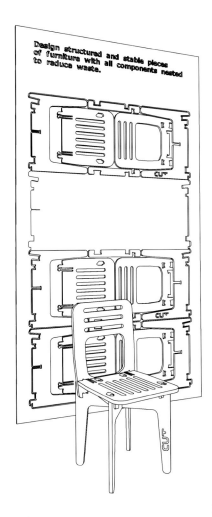

玛丽安娜·科斯塔·席尔瓦
椅子
制造，切割椅子
Red Valchromat®，丙烯酸清漆
250x145x1.6厘米（嵌板），38x44x81厘米（椅子）
制造：CUT FURNITURE
2012

Mariana Costa e Silva
Illustrative panel on chair manufacturing, and CUT furniture chair
Red Valchromat®, acrylic varnish finishing
250x145x1.6 cm (panel) | 38x44x81 cm (chair)
Production by CUT FURNITURE
2012

少

佩德罗·索托·梅弗

Less Bench

Pedro SottoMayor

设计师极力减少材料使用，尽可能的简洁。"少"这个名字也是对于现代思维的继承。通过移动金属面板上的小方块带来设计的动感和透明性。极简（Less）与极繁（More）形成鲜明对比。

Piece from the Less collection, consisting of three products for interior / exterior (two stools and a table). Pedro SottoMayor works on excess and the absolute necessity to focus on an economy of material resources. The title also puts us in the position of heirs of modern thinking, and the process of removing small squares from the metal sheet imparts dynamism and transparency to the design. The Less (part subtracted), is the More that makes the piece distinctive.

佩德罗·索托·梅弗
长凳
环氧粉末涂层金属，适合室内或户外使用
120x30x45厘米
制造：PSDI
2011

Pedro SottoMayor
Less Bench
Metal with epoxy powder coating for indoor or outdoor use
120x30x45 cm
Production by PSDI
2011

© Pedro SottoMayor

SIT'ABIT

PLY&co.（保罗·科斯塔）

SIT'ABIT

PLY&co. (Paulo Costa)

这款凳子因其环保和诙谐的设计理念显得与众不同。不需胶水或者螺丝即可轻松拆装，储存和运输也变得容易。它充分利用材料的潜力，具有较好的手感和坚固性。

A stool that stands out for its ecologically sustainable concept and for its amusing and playful spirit. Very easy to assemble, by simply fitting together (without glue or screws), and just as practical to store and transport. Making use of the materials' potential, it is both pleasant to the touch and very strong.

PLY&co.（保罗·科斯塔）
可拆卸凳子
桦木胶合板，压缩软木，无烟煤黑色
45x30x30厘米｜35x26x26 厘米｜45x30x30厘米
制造／出品：PLY&co.
2011/2012

PLY&co. (Paulo Costa)
SIT'ABIT birch, SIT'ABIT Jr. cork, SIT'ABIT Valchromat®
Demountable stools
Birch plywood, cork agglomerate, anthracite black
45x30x30 cm｜35x26x26 cm｜45x30x30 cm
Production/Edition PLY&co.
2011/2012

3. 大众与博识
Between the Popular and the Erudite

芒格
自然，色彩，软木
贡萨鲁·普鲁登修斯

Munge table and stools
Natural, Colour, Cork
Gonçalo Prudêncio

采用坚实松木、软木和涂料，"芒格"桌凳系列体现了设计师对工业化制造的经济和社会领域的批判性反思。采用来自葡萄牙的原材料，强调简洁大方。设计师出于对经济和环境的担忧，自创了"ecolomic"一词（代表生态和经济的融合），并围绕这一概念设计了一个家具系列。

In solid pine or in cork and lacquered versions, the Munge table and benches are part of Gonçalo Prudêncio's critical reflection on the social and economic aspects of industrial production. Made from Portuguese raw materials, they are notable for a formal brevity. Gonçalo Prudêncio's economic and environmental concerns have led him to invent the designation "ecolomic", (a fusion of the two concepts) and to develop a furniture collection entirely designed and produced by him (G.pt).

贡萨鲁·普鲁登修斯
桌子和凳子
木
桌子 | 76x60x60厘米
凳子 | 45x33x33厘米 | 33x 30x30厘米 | 33x30x30厘米
制造：Gonçalo Prudêncio Office for Design
2010

Gonçalo Prudêncio
Munge table and stools Natural, Colour, Cork
Wood
Table: 76x60x60 cm
Stools: 45x33x33 cm | 33x 30x30 cm | 33x30x30 cm
Production by Gonçalo Prudêncio Office for Design
2010

Nova Sogra

里卡多 · 德里扬和若 · 马蒂亚斯

这种中空的圆形小垫子叫做"小轮子"或者"继母",是过去妇女在头顶顶篮子或者水壶时用的垫子,通常用碎布、羊毛或者亚麻做成。今天,"小轮子"成为代表过去的装饰品。在这里,这一常见元素的创新改造,强化了传统手工艺品的经济和社会性。

里卡多 · 德里扬和若 · 马蒂亚斯
凳子
不锈钢,布条,刺绣线,绣花,编织
45x45x45厘米
制造: Nween
2011

Nova Sogra Stool

Ricardo Tralhão and João Matias

The "Little wheels" or "Mothers-in-Law" are small cushions, circular in form and open in the centre, traditionally used by women to carry baskets and water pitchers on their heads. They were made of braided and embroidered strips of rag, wool and linen. Today the "little wheel" is a decorative object, at most, representing the past. Here this popular element is re-used, and contributes to the economic and social enhancement of the craftsman's work.

Ricardo Tralhão and João Matias
Nova Sogra Stool
Stainless steel, rag strips, embroidery thread, embroided and braided
45x45x45 cm
Production by Nween
2011

布勒尔家具

贡萨鲁·坎波斯

Burel furniture seats

Gonçalo Campos

布勒尔使用葡萄牙羊毛手工制成。质地坚实，用途多样，可以通过各种质地和花色来展示。这意味着它可以有很多用途。近年来Burel更多地将传统制作与现代设计相结合。设计师在设计这款凳子时将布勒尔作为内衬，并赋予它更多舒适度。

Burel is an artisanal Portuguese fabric made entirely of wool. Very strong and versatile, it can take on a vast range of textures, patterns and colours. These characteristics mean that it can be used for different purposes. In recent years burel has undergone a process of modernisation, combining more contemporary design with traditional manufacturing processes. In this stool by Gonçalo Campos, burel is used as an upholstery material imbuing it with greater comfort.

贡萨鲁·坎波斯
椅子
41种颜色（两种颜色或一个单一的颜色组合）
53x47厘米（直径）
制造：Burel Factory
2011

Gonçalo Campos
Burel furniture seats
Burel
41 colours (combination of two colours or a single colour)
Ø53x47
Production by Burel Factory
2011

任何

Pedrita（丽塔 · 若和佩德罗 · 费雷拉）

Any carpets

Studio Pedrita

遵循葡萄牙传统钩编地毯从不浪费线的原则，该设计是可持续性理念的拥护者。在半手工制作过程中使用一般制作中废弃的羊毛，最大限度减少废弃物。造型为圆形小地毯，色泽对比大胆，带有明显的Pedrita设计工作室的风格。大众材料文化和传统葡萄牙技艺是设计的重要灵感。

Following the tradition of Portuguese crocheted blankets where no thread is wasted, the Any carpets are an endorsement of sustainability. The waste wool from normal manufacturing was reintroduced in a semi artisanal manufacturing process that minimizes waste. The result is circular rugs with a strong colour and visual identity For the Pedritas, popular material culture and traditional Portuguese techniques are an important source of inspiration.

Pedrita（丽塔 · 若和佩德罗 · 费雷拉）
地毯
羊毛
70厘米（直径）
制造：Piodão Group
2011

Studio Pedrita (Rita João and Pedro Ferreira)
Any carpets
Wool
70cm diameter
Production by Piodão Group
2011

三张桌凳

门德斯和马赛多

Three Table Bench

Mendes & Macedo for Galula

"三张桌凳"有三个面和三条腿,可以通过很多方式装配,可以用作凳子或者边桌,展示了软木的创新利用。

Três Stool/ Table has 3 sides, 3 legs, but many ways of fitting together, and adapts to any person and posture. It acts as a stool or as a modular side table, and is one of the new ways of working with cork.

门德斯和马赛多
桌子和凳子
40.8x35.5x44厘米
制造:Galula
2012

Mendes & Macedo for Galula
Three Table Bench
High Density Cork, thermo-coated metal
40.8x35.5x44 cm
Production by Galula
2012

储物柜

丹尼尔·杜阿尔特

Apothecary Cabinet

Daniel Duarte

这是对17世纪药剂柜和已不再使用的家具样式的重新诠释。它由老橡木、黄铜和陶瓷材料制成，呈现出古董的色泽。多抽屉的设计寓意着柜子里隐藏着的秘密，体现古代药剂柜的神秘感。

A reinterpretation of 17th century apothecary cabinets and of a furniture typology currently in disuse. This piece is built from aged oak, brass and ceramic material so as to take on an antique aspect. The numerous drawers suggest the existence of various secrets, as happens with antique apothecary cabinets.

丹尼尔·杜阿尔特
储物柜
中密度纤维板和橡木
103.5 x 152 x 33.2厘米
2012年椅子上的艺术参展作品
制造：Jocilma, Indústria de Móveis S.A.
2012

Daniel Duarte
Apothecary Cabinet
MDF and oak
103.5 x 152 x 33.2 cm
Design developed in the context of Art on Chairs 2012
Manufacturer: Jocilma, Indústria de Móveis S.A.
2012

多
罗德里戈·维瑞内斯

Multi
Rodrigo Vairinhos

突破所有传统概念的限制，MULTI是边桌、床头柜和书籍的集合体。它可以放书、杂志、CD或者DVD，可以在家里的任何角落使用。

Unconstrained by any conventions, MULTI can be considered a fusion between a side table/ bedside table, and shelves. As well as holding books, magazines, CDs or DVDs, it is a piece that can be used anywhere in the domestic environment.

罗德里戈·维瑞内斯
多功能家具
山毛榉木
40x57.5x37厘米
制造：罗德里戈·维瑞内斯
2012

Rodrigo Vairinhos
Multi
Multipurpose piece of furniture
Beech wood
40x57.5x37 cm
Production by Rodrigo Vairinhos
2012

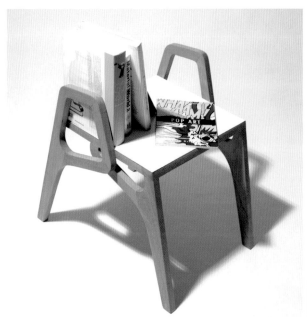

皮尔逊
米格尔·里奥斯

Pirson
Miguel Rios

米格尔·里奥斯认为,椅子是代表思考的物体。设计中融合唐纳德·贾德的美学极简主义,让·弗朗索瓦·皮尔逊的建筑分析以及路易斯·阿尔伯特·内科尔的临界实验,引发人们对于艺术和设计、形式和功能、真实和虚构、感觉和实物之间的二元关系。

The chair as an object of contemplation and reflection, this is Miguel Rios' proposal. Crossing references with the aesthetic minimalism of Donald Judd and the architectonic discourse of Jean François Pirson and the critical experiments of Louis Albert Necker, triggering a conceptual reflection on the dualities of art and design, form and function, real and fictional, perception and concrete matter.

米格尔·里奥斯
椅子
橡木
45x45x90厘米
出品: Miguel Rios Design
2010

Miguel Rios
Pirson : "Je pourrais aussi regarder la chaise comme un object (parmi d'autres)..."
Oak wood
45x45x90 cm/ each
Edition by Miguel Rios Design
2010

4. 这不(仅仅)是个烟斗
This is not (just) a Pipe

花瓶台灯
贡萨鲁·坎波斯

Vase Lamp
Gonçalo Campos

本设计结合了家中常见两个物件台灯和花瓶。设计师旨在探寻每个物件的诗意和传播意义，使用精炼的幽默和直观的智慧。它让人微笑和思考，它的简约和荒诞令人惊讶。

This lamp combines two of the most common objects in household use: the vase and the lamp. Gonçalo Campos has come to explore the poetic and communicative meaning of each object, using refined humour and an intuitive intelligence. This piece is an example of this position. It makes us smile and think, and surprises us with its simplicity and a certain nonsensicality.

贡萨鲁·坎波斯
台灯
陶瓷
50 x 40 x 120 厘米
制造：Show me Design & Art Gallery
2012

Gonçalo Campos
Vase Lamp
Ceramic
50 x 40 x 120 cm
Manufacturer: Show me Design & Art Gallery
2012

葡萄牙马鞍
保罗·帕拉

Portuguese saddle
Paulo Parra

从葡萄牙斗牛文化和传统中汲取灵感,这一作品是一件充满趣味性的多用途物件。由阿连特茹地区的工匠使用该地区传统材料软木和皮革制成。这些传统材料和马鞍造型的运用,让坐在上面的人得到一种舒适、自然、安静和有趣的体验。

Drawing on the culture and tradition of Portuguese bull fighting, the Sela is a multi-purpose object that invites playfulness. Produced by craftsmen in the Alentejo, it is made from the traditional materials of this region: cork and leather. These traditional materials, allied to the form of the saddle, allow one to sit naturally, comfortably, smoothly and silently but at the same time, invite fun and games.

保罗·帕拉
座椅
压缩软木
45x40x50厘米
出品:OficinaOxigénio
1996

Paulo Parra
Portuguese saddle
Cork agglomerate
45x40x50cm
Edition by OficinaOxigénio
1996

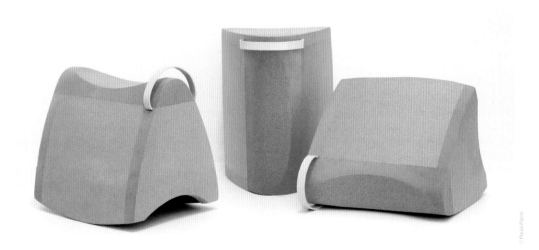

达达尼昂
托尼·格里洛

Dartagnan Armchair
Toni Grilo

以最简约抽象的方式表明椅子的本质：橡木框架间悬置的一条皮带代表坐的位置。元素化的设计风格带来了极强的视觉张力，让人不禁反思椅子的视觉原型和它在西方材料文化中的地位。

Extremely minimalist, the function of a chair is reduced to the essential: A simple strip of leather hung from an oak structure suggests the place of the seat. This gesture, despite being almost elemental, results in a piece of great visual strength and forces a reflection on the archetype of the chair and its place in western material culture.

托尼·格里洛
扶手椅
自然色或黑色橡木，黑色皮革
75x60x73厘米
出品： Haymann
2012

Toni Grilo
Dartagnan Armchair
Natural or black oak with black leather
75x60x73 cm
Edition by Haymann
2012

目标

费尔南多·布瑞斯奥

Target table

Fernando Brízio

费尔南多·布瑞斯奥设计的简约来自于对材料文化的成熟、创意理解和各种材料实验。设计师突破物件的实用效用，探寻其诗意和传播价值。这件设计表现了在射中靶子之前的情节，具有明显的原创性、不可预知性和讽刺意味。

The simplicity of Fernando Brizio's designs result from a mature and creative conceptual process on material culture and from experimenting with materials. Going beyond the practical utility of each object, he works on the concepts and the possible levels of significance/meaning of each object, exploring its poetics and communicational value. This piece suggests an action prior to shooting the target and is notable for its originality, unpredictability and irony.

费尔南多·布瑞斯奥
桌子
木和箭
50x45（直径）
时间间隔展览参展作品
出品：费尔南多·布瑞斯奥
2009

Fernando Brízio
Target table
Wood and arrows
ø50x45 cm
Project presented at the exhibition Lapse in Time
Edition by Fernando Brizio
2009

手柄

费尔南多·布瑞斯奥

Handle stools

Fernando Brízio

费尔南多·布瑞斯奥的设计总是让欣赏者经历质疑、想象、反思和最终的会心一笑。而这源于一种严格的勤勉和由此产生的解密过程。凳子边缘加的把手，完全颠覆了凳子的传统存放和放置方式，可以将它挂在墙上。

Fernando Brizio's designs arouse in the viewer questioning, imagination, reflection and a smile. This results from rigorous formal diligence and the actual deciphering process triggered by it. In this case, the addition of a simple handle to a stool means that the usual way of handling or storing it is radically altered, so that it can now be hung or fixed to the wall.

费尔南多·布瑞斯奥
凳子
木和毡布
44x30x30厘米
里斯本系列
制造：Tema Home
2009

Fernando Brízio
Handle stools
Wood, felt
44x30x30cm
Lisbon Collection
Production by Tema Home
2009

Hole Cups

Fernando Brízio

This piece seeks to establish a personal and emotional connection with the user. Illustrative of the importance attributed by Fernando Brízio to the design and performative character that many of his pieces possess, this object is only completed by the action of its user. The lines of colour that may or may not be traced on the surface are an integral part of the piece.

Fernando Brízio
Hole Cups
Cork agglomerate, colouring pencils
D255x28cm
Production by MATERIA Cork by Amorim, curated by Experimenta Design
2011

软木队长吊灯

米格尔·阿鲁达

Captain Cork Mini / Maxi Chandeliers

Miguel Arruda

这一灯组将自然材料引入家居环境，充分探索软木隔热和隔声的特性。轻便的组合赋予其多元和轻松的使用体验，可以置于等肩的高度，作为局部照明使用，或者直接放在桌上。尺寸和形式恰到好处，可以让人轻松地操作。

The light fixtures Captain Cork bring a natural material to the domestic environment – cork – and explore its qualities of thermal and acoustic insulation. The result is light fixtures that can be put highly diverse and informal uses. They can be put above shoulder height for more localised illumination, or simply be placed on a table. Their size and format also allow them to be easily manipulated.

米格尔·阿鲁达
吊灯
压缩软木
13x23厘米（直径）
出品：Dark
2013

Miguel Arruda
Captain Cork Mini / Maxi Chandeliers
Cork agglomerate
Ø13x23 cm
Edition by Dark
2013

推车
琼娜·卡布瑞塔·马丁斯

Cart
Joana Cabrita Martins

这辆小车几乎全部使用现成物品构成,自行车轮、水果木条箱和雨伞手柄按照设计师的想象结合在一起。过时和毫无价值的材料由此被赋予了新的生命和尊严。除了实用性之外,它的存在也使我们以更具创造性的眼光看待周围的事物。

Almost a ready-made, this cart is the result of the meeting of imagination and materials as varied as bicycle wheels, fruit crates and umbrella handles. Banal and unvalued materials take on a new life and dignity. In addition to its utilitarian meaning, its existence makes us look with more creativity at the world of objects around us.

琼娜·卡布瑞塔·马丁斯
推车
自行车轮,水果箱,雨伞柄
87x45x76厘米
制造/出品:Remix Project
2014

Joana Cabrita Martins
Cart
Bicycle wheels, fruit boxes, umbrella handles
87x45x76 cm
Production/Edition by Remix Project
2014

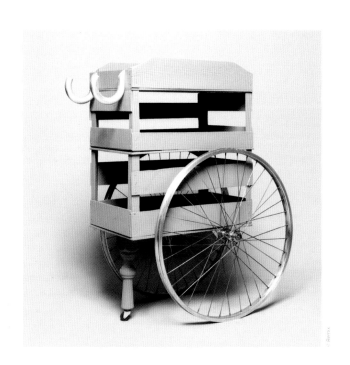

桌子
安德烈·卡拉多

Table
André Calado

本设计属于"重混"项目的一个部分,该项目发起的目的在于倡导民众与社区之间建立更紧密地联系,从而促进社会融合和发展。从环保角度出发,瓷盘、木块、塑料、织物、木板箱等都被赋予新的功能,共同构成全新的设计系列。

A product made as part of Project Remix, an initiative for social inclusion and development based on civic and community participation. Reusing materials and working from an ecological angle, everything serves as a basis for design. Ceramic plates, pallets, wood blocks, blinds, plastics, fabrics and fruit cartons take on new functions, and originate a new collection of objects.

安德烈·卡拉多
桌子
木托盘
30x45x120厘米
制造/出品:Remix Project
2012

André Calado
Table
Wood Pallets
30x45x120cm
Production/Edition by Remix Project
2012

书桌

米格尔·维埃拉·巴普蒂斯塔

Book table

Miguel Vieira Baptista

一本打开的书构成了小边桌的基础和本质。书和文字的暗示既创造一种阅读的环境，又赋予物件严格效用的一种新的解读。

The pages of an open book are the basis and essence of a small side table. At the same time, the allusion to a book and the written word create a wide universe of readings and interpretations that go beyond the strict utility of the object.

米格尔·维埃拉·巴普蒂斯塔
桌子
60（高）x 55（直径）厘米
出品：Cristina Guerra Contemporary Art
2005

Miguel Vieira Baptista
Book table
Paper, fabric, glass
h60 x D55cm
Edition by Cristina Guerra Contemporary Art
2005

图书在版编目（CIP）数据

椅子上的艺术 /《casa国际家居》杂志编. -- 北京：
新星出版社，2014.9
 ISBN 978-7-5133-1610-1

Ⅰ.①椅… Ⅱ.①c… Ⅲ.①椅－设计 Ⅳ.
①TS665.4

中国版本图书馆CIP数据核字(2014)第207648号

椅子上的艺术
《casa国际家居》杂志编

选题策划：《casa国际家居》杂志
网　　址：www.casainternationalmag.com
责任编辑：汪　欣
装帧设计：马塔·博尔赫斯（葡萄牙）
出版发行：新星出版社
出　版　人：谢　刚
社　　址：北京市西城区车公庄大街丙3号楼100044
网　　址：www.newstarpress.com
电　　话：010-88310888
传　　真：010-65270449
法律顾问：北京市大成律师事务所
读者服务：010-64563942 info@casainternationalmag.com
邮购地址：北京市西城区车公庄大街丙3号楼100044
印　　刷：北京尚唐印刷包装有限公司
开　　本：889mm×1194mm　1/16
印　　张：11
字　　数：16千字
版　　次：2014年9月第一版　2014年9月第一次印刷
书　　号：ISBN 978-7-5133-1610-1
定　　价：180.00元